Caregiving
Research • Practice • Policy

Ronda C. Talley, Series Editor

An official publication of
The Rosalynn Carter Institute for Caregiving

For other titles published in this series, go to
www.springer.com/series/8274

Seymour B. Sarason

Centers for Ending

The Coming Crisis in the Care of Aged People

Foreword by Saul B. Cohen

🐴 Springer

MT

Seymour B. Sarason (deceased)
Department of Psychology
Yale University
New Haven, CT 06520

ISBN 978-1-4419-5724-5   e-ISBN 978-1-4419-5725-2
DOI 10.1007/978-1-4419-5725-2
Springer New York Dordrecht Heidelberg London

Library of Congress Control Number: 2011930251

Printed on acid-free paper

Springer is part of Springer Science+Business Media (www.springer.com)

6/6/14

To Lisa
Who deserves more than mere words of
gratitude and affection

Series Foreword

From its inception in 1987, the Rosalynn Carter Institute for Caregiving (RCI) has sought to bring attention to the extraordinary contributions made by caregivers to their loved ones. I grew up in a home that was regularly transformed into a caregiving household when members of my family became seriously ill, disabled or frail with age, so my interest in the issue is personal. In my hometown of Plains, Georgia, as in most communities across our country, it was expected that family members and neighbors would take on the responsibility of providing care whenever illness struck close to home. Delivering such care with the love, respect, and attention it deserves is both labor-intensive and personally demanding. Those who do so represent one of this nation's most significant yet underappreciated assets in our health delivery system.

When the RCI began, "caregiving" was found nowhere in the nation's health lexicon. Its existence was not a secret but rather simply accepted as a fact of life. In deciding on the direction and priorities of the new institute, we convened groups of family and professional caregivers from around the region to tell their personal stories. As I listened to neighbors describe caring for aged and/or chronically ill or disabled family members, I recognized that their experiences reflected mine. They testified that while caregiving for them was full of personal meaning and significance and could be extremely rewarding, it could also be fraught with anxiety, stress, and feelings of isolation. Many felt unprepared and most were overwhelmed at times. A critical issue in the "field" of caregiving, I realized, was the need to better understand the kinds of policies and programs necessary to support those who quietly and consistently care for loved ones.

With the aging of America's Baby Boomers expecting to double the elderly population in the next 20 years, deinstitutionalization of individuals with chronic mental illnesses and developmental disabilities, a rising percentage of women in the workforce, smaller and more dispersed families, changes in the role of hospitals, and a range of other factors, caregiving has become one of the most significant issues of our time. Caregiving as an area of research, as a focus and concern of policy making, and as an area of professional training and practice has reached a new and unparalleled level of importance in our society and indeed globally.

As we survey the field of caregiving today, we now recognize that it is an essential component of long-term care in the community, yet also a potential health risk for those who provide care. The basic features of a public health approach have emerged: a focus on populations of caregivers and recipients, tracking and surveillance of health risks, understanding the factors associated with risk status, and the development and testing of the effectiveness of various interventions to maximize benefits for both the recipients of care and their providers.

The accumulated wisdom from this work is represented in the volumes that make up the Springer Caregiving Series. This series presents a broad portrait of the nature of caregiving in the United States in the twenty-first century. Most Americans have been, are now, or will be caregivers. With our society's increasing demands for care, we cannot expect a high quality of life for our seniors and others living with limitations due to illness or disability unless we understand and support the work of caregivers. Without thoughtful planning, intelligent policies, and sensitive interventions, there is the risk that the work of family, paraprofessional and professional caregivers will become intolerably difficult and burdensome. We cannot let this happen.

I am pleased to include within this series Dr. Seymour Sarason's very personal reflections on being the recipient of care. His volume, *Centers for Ending: The Coming Crisis in the Care of Aged People* describes the process through the eyes and heart of the care recipient. Combining personal reflections and policy insights, Sarason sensitizes readers to appreciate caregiving in terms of both its form and its emotional tone through the eyes of its recipient. In that sense, the volume differs from that of other volumes in the series but it does so in a complementary manner. In its essence, Sarason's *Centers for Ending* reminds us that caregiving is an intimately interpersonal and reciprocal process involving both those who deliver and those who receive care.

Readers of this series will find hope and evidence that improved support for family and professional caregivers lies within our reach. The field of caregiving has matured and, as evidenced in these volumes, has generated rigorous and practical research findings to guide effective and enlightened policy and program options. My hope is that these volumes will play an important role in documenting the research base, guiding practice, and moving our nation toward effective polices to support all of America's caregivers.

Rosalynn Carter

Foreword

Centers for Endings: The Coming Crisis in the Care of Aged People is a rallying call to action that comes from knowledge, from experience, and from the heart. Revered psychologist, Seymour Sarason, builds on years of association with both the medical and educational establishments, plus recent personal experience, to describe and diagnose the problems and prescribe remedies.

Policy makers can no longer ignore the demographic reality of an explosion of elderly in the population, combined with longer life spans and dispersion of families, resulting in the urgent need for elder care in various forms. But fulfilling the need for more facilities will not answer the woeful deficiencies in quality of care in even the better nursing homes, let alone the publicly supported and business-focused institutions. Dr. Sarason zeroes in on the personnel responsible for the most constant and intimate daily care, the lowest rung of the hierarchy, the aides, whom he also considers victims of a dysfunctional system. He calls for focusing on drastic improvement in their selection, supervision, salaries, and, especially, training. All of which comes at a high cost.

Dr. Sarason does not shirk at addressing further economic realities. On the demand side, it is a small percentage of the population that can afford today's cost for the better senior residence centers and nursing homes. He provides dollar and cents prices for such facilities, demonstrating that the incomes of most middle- and working-class retirees cannot match these costs for any extended period. This speaks to governmental involvement, for which he makes specific recommendations, such as the convening of a presidential commission. The problems have been well studied, and the commission should focus on action. Citing previous national solutions to great social and health challenges, such as social security, the GI Bill, the Veterans Administration, and Medicare/Medicaid legislation, he issues a clarion call for action to address the looming social and economic crisis of the aging "baby-boom" generation. This will require the kind of political will and leadership that has been lacking so far.

However, beyond quality medical care and physical facilities, it is the approach to the elder resident or patient as a thinking, feeling, sensitive human being about which Dr. Sarason is most passionate. He is particularly sensitive to the loneliness and fears of those facing life's end. His writings as a clinical psychologist have long dealt with the need for caring and compassion in many social service settings. Now he shares deeply personal feelings and experiences resulting from his own stays in hospitals, a senior residence and nursing homes. He decries the attitude of caregivers in their lack of interpersonal relationships with elderly patients, dealing with them as objects rather than individuals who deserve consideration, consultation, and dignity.

Every circumstance and phenomenon that the author highlights resonates personally. My sister is resident in a well-regarded nursing home, yet often experiences the indignities and lack of personal concern on the part of staff, as well as the roadblocks of institutional bureaucracy. In addition, as a member of the New York State Board of Regents, I serve on the review committee of its Office of the Professions. An inordinate number of the cases which come before us involve malfeasance in hospital, nursing, and home care settings. While the upper echelons of the hierarchy require certification, much of what goes wrong relates to their lack of communication with the patient, as well as other deficiencies. With aides, this is magnified by their casual recruitment and selection, lack of standards, often poor language skills, absence of training, and low pay scale.

This valuable volume is comprehensive in its coverage of the roots of the problems, their manifestations, and direction of solutions. It should be required reading for medical personnel, social workers, operators of senior facilities, and policy makers. Individuals facing the later years and their families will be enlightened and moved, and better prepared to cope with the decisions and challenges of aging.

President Emeritus   Saul B. Cohen
Queens College
CUNY
Flushing, New York

Contents

Chapter 1
Themes of the Book

I never expected to write this very personal book about aged people and nursing homes. My late wife and I had extensive experience with these facilities in connection with our parents in the 1960s and 70s. These were sobering experiences for several reasons. The first was that what passes for care of old people was impersonal and devoid of any attempt to keep the minds of residents active, to prevent retreat into a private world that stayed private until they died. Their "bodies" were never neglected, but their minds were.

The second reason for my sobering reactions was financial. My father fell a couple of times and no one was in the room to see or hear his first cries. My wife and I decided to hire a practical nurse who would be with him during the day 5 days a week. We were fortunate to locate a person who, as my wife said, was Italian, but undoubtedly had all of the characteristics of the stereotypic Jewish mother and grandmother. She read his beloved New York Times to him (his eyesight was poor), had long discussions with him, took him in a wheelchair for walks in and out of the facility, and made sure he ate every bit of every meal. He lived contentedly for another 2 years. That cost us $21,000 for each of the 2 years. Today, the cost would be double that amount. That is when I realized that there was a two-tiered system: the haves and the have-nots. I was happy to be able to afford the financial sacrifices for my father's care, but I also felt guilt that he was the only one in that nursing home to get the care that everyone else there needed but did not get.

Hospitals and nursing homes are different kinds of places, but they have much in common. In the last chapter of *Caring and Compassion in Clinical Practice* (Sarason, 1986), I had written about my experiences to that point in my life with hospitals and nursing facilities. In writing that chapter, I did not know how significantly treatment and extended care settings would affect my life in the years ahead. In 1993, my wife and I were in a serious auto accident. My wife was killed instantly and I spent the next 3 months in three different hospitals, one of which specialized in rehabilitation. Overwhelmed by the loss of my wife, I was only partially aware of my treatment and rehabilitation experiences during those emotionally and physically painful months. Fifteen years later, those traces emerged vividly! Having moved to a retirement community some years before, I fell, fractured my hip and pelvis, and had to enter the nursing care facility within the retirement community.

S.B. Sarason, *Centers for Ending*, Caregiving: Research, Practice, Policy,
DOI 10.1007/978-1-4419-5725-2_1, © Springer Science+Business Media, LLC 2011

That traumatic experiences that marked my stay in that facility was one reason that I wrote this book. The larger reason, however, was that it made me painfully (literally as well as figuratively) aware of an impending crisis. America has not yet taken seriously an inescapable fact: We are at the beginning of a dramatic increase in the number of aged people. We are unprepared for that brute demographic fact which will transform our society. With those millions of elderly citizens will come needs for medical and nursing care that have yet to be recognized, planned for and begun! Sadly, I believe that it will only be acknowledged after far too many individuals as well as families have suffered.

I expand on these and associated issues in the pages of this book. Initially I was reluctant to go into details about my story in the nursing home because it might convey the impression that I was a complaining, egocentric, old fogy unwilling or unable to tolerate frustration. I don't think of myself as such, but even if I were that would not negate the implications of my experiences for the millions of citizens nearing retirement and old age. Shortly after I entered the nursing care facility within my retirement community, I soon learned from other residents that there was no point in "fighting city hall" even if your condition was atypically extreme: I have one dead weight polio arm, the other arm showing all the symptoms of the post polio syndrome. I am legally blind; I have hearing loss. For more than 5 weeks, I was not permitted to stand on my feet. Transferring me from my bed to a chair required a contraption called a hoist, plus the assistance of two aides. I was not, to indulge in understatement, a happy camper in a camp in which most residents had resigned themselves to a life of privacy and loneliness. It brought back memories of my father's experience as a nursing home patient in his final year and of my experiences in three hospitals recuperating from that terrible auto accident.

After a week in the nursing home, I gave thought to writing this book. As noted, the decisive factor in planning this volume was the realization that in a few decades, millions of baby boomers born during and after World War II would contain many people who toward the end of their lives would need or want to live in a total care facility that contained a nursing home. Who was giving thought to that future? It is not only a matter of building facilities. Although there is already a shortage of facilities, especially for the have-nots, that is not the most important problem until as a society we get clear agreement about how to answer the moral question: What do we owe any sick, frail aged person who will live his or her final years in these facilities? And in confronting that question, we will run into another equally thorny question: How do we select and train personnel at all levels of responsibility to understand far better than they now do the phenomenology of aged people?

Let us take the obvious seriously: Personnel who take care of such people have never been aged! Indeed, in the course of their lives these personnel have acquired stereotypes that purport to "explain" the behavior, overt and covert, of old people, most of which is wrong or demeaning, illogical, mindless, or insensitive. How many readers of this book who are parents have found themselves saying that their experience of parenthood gave them a better appreciation of the behavior of their parents? How many times in the seminars I taught did I discuss the Great Depression, knowing that the students had no way to understand me when I would say that the Great

Depression forever scarred me? It is like trying to tell someone who has never fought in a war what war is like.

The goal in selection and training is not the utopian one of perfection. We will always fall short of the mark. We can live with that but we must not allow ourselves to live with a state of affairs which make for human warehouses. We cannot allow ourselves to be ignorant of the history of public institutions for the mentally ill, the mentally retarded, juvenile delinquents, and the aged poor. They were all warehouses, which is what many nursing homes are today. And in all of them the personnel who spent the most time with residents were unselected, poorly trained, had the least education, and were pitifully paid. It is not much different today.

Will it be any different for the future aged people? The answer is no and for several reasons. The first is that it is already the case that the fastest growing age cohorts are those above seventy. The second is that cohort will increase exponentially in the next few decades. The third is that our public officialdom—from the President on down—cannot be more ignorant and silent about what is coming down the road, and the fourth is that time is not on our side.

But hope does spring eternal. My hope is that a concatenation of events leads a future president to appoint a commission or task force to study and make recommendations for the care of aged people. The second half of this book is about four such reports which, unlike most official reports, did not gather dust in the national archives. Why was it different in the case of these reports? I discuss the social security act in the mid 1930s, the GI Bill of Rights at the end of World War II, the Head Start legislation in 1965, and the 9/11 report on the unpreparedness of government for 9/11. I end the book with some guidelines and concrete ideas for a commission on aged people. Going back over these four reports was very instructive. I hope it will be for the readers of this book. We can learn from the past, but only if we search it with respect and specific purpose.

I am not aware that anyone has written a book on the history of national commissions, especially in regard to their practical consequences. If a future president appoints one to study the issues I raise in this book, I hope he or she has the good sense to become aware of one feature of the 9/11 commission whose object of study was the culture of government in Washington. It was not the culture of government in New York or any other big, complicated setting. It was the structure and dynamics of the national government in Washington, DC. *Every member of the commission had spent years in that government in one or another capacity or relationship. They knew how it worked or did not work, and why. They knew what went on in public and behind the scenes, the difference between appearance and reality, between principled partisanship and party allegiance. And they knew that in the context of the bombing of the Twin Towers the search for truth took precedence over everything and everyone else.* That explains why the last chapter of this book where I discuss the selection of members for a commission on the aged is titled "Guidelines and Caveats."

President Richard Nixon's vice president once said, "When you see one slum, you have seen them all." I do not know what Agnew meant by "see." You cannot see the experience of a slum's residents. You cannot see how that experience is different with different age cohorts. You cannot see how and why living in a slum sustains and reinforces dependency and resignation, and a sense of worthlessness. Seeing a slum is easy. Understanding it is very different and difficult.

Residential facilities for aged people are not slums, even those primarily serving poor people for whom the state reimburses the facility. I should hasten to add, however, that federal and state health agencies find many nursing homes below standards of care, and those infractions make grim reading and far more often than not became a statistic because of a family member or friend, or anonymous whistleblowers.

The voices of the aged are not sought or tabulated. Aged residents are perceived as "on their way out" and are expected to go out with conforming grace. Whatever you may mean by human care of the aged should refer to more than keeping them alive.[1] As important as that is, it should refer to *feeling* alive. If you regard that as mushy and sentimental, it is because you have yet to become aged and cannot grasp that aged people think and feel and want human contact and relationships and do not want to resign themselves to an unwanted privacy because those who care for them show no signs of wanting to hear. The world of the aged and the world of the caretakers are two different phenomenologies. They should not be so totally different.

That is why I emphasize that the members appointed to the commission should not be chosen because of their public status or the labels of their profession. And in the later discussion of a commission I give a concrete example of someone who should be on the commission. He is, of all things, an aged pediatrician!

We are used to hearing that aging and dying is a normal process during which we are beginning to fall apart at the seams. No one is exempt from that process. The implication is that we should not fear it, we should resign ourselves to the inevitable, to a universal fate. And this resigning comes with a description, very objective, of why the bodily seams are coming apart and cause death. What such a view ignores or glosses over is that the human brain is part of the body and for most people it continues to be active. It fantasizes, has needs for the presence of others who will dilute or prevent the stabbing sense of aloneness and loneliness, personnel who will try by actions and words to convey that they have a sense of what he or she is experiencing, but has trouble putting into words. Presence and touch are the major medications. These are not medications dispensed in nursing homes. Yet, as noted by Spitz six decades ago, their absence at the beginning of life can have serious negative emotional and physical consequences. In infants, these consequences raised concerns. Might their appearance in the elderly be overlooked, or worse, misattributed to the aging process? If that were to occur as the tidal wave of the elderly begins to crash on the unprepared shores of our nation's nursing homes, we may subsequently re-discover an epidemic of anaclitic depression in the elderly

---

[1] Readers are reminded of Rene Spitz's (1945, 1946) seminal work on the nature and causes of anaclitic depression or hospitalism in institutionalized infants. This diagnosis was used in the 1930s to describe infants who wasted away while in hospital. Spitz' research linked the presenting cognitive and physical losses to a lack of social contact between the infant and its caregivers. Might nursing care responsive to the physical needs of the elderly have equally negative iatrogenic effects?

and find too late for too many that a bit of attention and social interaction could have made a world of difference!

In reading this book, the reader may justifiably find him or herself asking: What permits the author to use personal experiences to make generalizations about aged people, nursing homes, hospitals, total care facilities, nurses, physicians, and institutional administrations? That is a legitimate question and my answers are several. The first answer is that I make it clear that I did not come to live in a total care facility to study that facility and the people and personnel in it. I was not interested in studying or proving anything. I was too preoccupied with my own personal and medical problems to give thought to anything that could be dignified by labeling it a study. You do not have to be a psychologist to know and expect that people vary widely and wildly on any human attribute, especially under conditions of stress. When at an advanced age you feel compelled to move to a total care facility where you will feel safe, protected, and cared for, and where you will end your days, you know what stress is and the toll that it takes before, during, and after you have made the fateful decision and move. Once you have "settled in" you have already concluded that you have or will contribute to the heterogeneity of those who are residents, a fact I emphasize and illustrate in this book. However, as the weeks and months pass, you are unaware that you have come to conclusions or generalizations masked by the surface heterogeneity. Call them what you will, these generalizations are based not only on what people say, but also on thoughts and feelings they infrequently can bring themselves to verbalize. For example, I say in later pages that not a day passed that I was not reminded of my infirmities, fears, and an uncertain, foreshortened future. A resident died or someone was taken to the hospital or to the nursing home or now needed a wheelchair or had fallen or someone's speech day by day became more unintelligible because of Parkinson's disease. These were all *external* reminders. I had internal ones as well, but there was one that intruded many times during each day: Will my macular degeneration get to the point where I would be unable to write? There are many residents with macular degeneration and I feel justified in saying that I know what they are reminded of every day. Phenomenologically, feeling alone is very different from the feeling of loneliness and it is the poignancy and fear of loneliness that haunts every resident, especially when you know you are near death. The last thing a resident wants to talk about is loneliness. It is enough that I think about it.

I think it was legitimate to question the validity of generalizing from my personal experiences. That is why I devote a chapter to proposing that a presidential commission be formed to study and assess the moral, medical, psychological, economic aspects of facilities and programs for aged people (numbering in the millions) who will need institutional services of various kinds. Our knowledge base is, to say the least, inadequate. Selection and training of caregivers is hardly in its infancy. The economic costs will be huge. The only group who can say, "We told you so," is the demographers who shortly after World War II had the data, wrote about it, and no one listened.

I cannot deny that I came to the center not expecting all would be sweetness and light. By virtue of a long life and a career in psychology, I know that human services institutions leave much to be desired, a few exceptions aside. I was mightily impressed in 1946 by a book by Albert Deutsch, *The Shame of Our State*, which described the treatment of patients in state hospitals. From 1942 to 1945, I visited numerous institutions whose missions were to educate and care for mentally retarded children. They were not pleasant experiences; in each instance, they were periodically publicly exposed as cases of inequity. When the Medicare legislation was passed in 1964, a few years later the mass media was reporting in gory detail nursing home scandals in various cities.

In 2007, a surgeon on the staff of the Harvard Medical School wrote a paper which confirms some of the major themes of this book (Gawande, 2007). Here are some excerpts.

> The geriatrics clinic—or, as my hospital calls it, the Center for Older Adult Health—is only one floor below my surgery clinic. I pass by it almost every day, and I can't remember ever giving it a moment's thought. One morning, however, I wandered downstairs and, with the permission of the patients, sat in on a few visits with Juergen Bludau, the chief geriatrician. . .[one patient] said that her internist had recommended that she come.
>
> About anything in particular? the doctor asked.
>
> The answer, it seemed, was yes and no. The first thing she mentioned was a lower-back pain that she'd had for months. . . .She also had bad arthritis. . .She'd had both knees replaced a decade earlier. She had high blood pressure "from stress,". . .She had glaucoma. . .She'd also had surgery for colon cancer and, by the way, now had a lung nodule that the radiology report said could be a metastasis—a biopsy was recommended.
>
> Bludau needed to triage by zeroing in on either the most potentially life-threatening problem (the possible metastasis) or the problem that bothered her the most (the back pain). But this was evidently not what he thought. . .
>
> Whether we admit it or not, most doctors don't like taking care of the elderly. "Mainstream doctors are turned off by geriatrics, and that's because they do not have the faculties to cope with the Old Crock," Felix Silverstone, the geriatrician told me. "The Old Crock is deaf. The Old Crock has poor vision. The Old Crock's memory might be somewhat impaired. With the Old Crock, you have to slow down, because he asks you to repeat what you are saying or asking. And the Old Crock doesn't just have a chief complaint—the Old Crock has fifteen chief complaints. How in the world are you going to cope with all of them? You're overwhelmed. Besides, he's had a number of these things for fifty years or so. You're not going to cure something he's had for fifty years. He has high blood pressure. He has diabetes. He has arthritis. There's nothing glamorous about taking care of any of those things."

I hope and expect that by the time the reader finishes this book he or she will have no doubt that the number of elderly persons in the population is rapidly growing and that the rate of that increase will also increase in the next few decades. I also expect that they will be both shocked and surprised by two facts. Dr. Gawande[2] reports: The first fact is that between 1998 and 2004, the number of certified geriatricians *fell* by one-third. The second fact is that during the same period applications to training programs plummeted while fields such as plastic surgery and radiology had record numbers of applications.

---

[2]On May 24, 2007, Dr. Gawande wrote a guest column for the New York Times in which he pinpoints some of the central issues I take up in this book. I know of no geriatrician who has written about these issues.

During the 2 months I was in the center's nursing home, I know I was regarded as an Old Crock, but not because I expected them to do more about my broken hip and pelvis. I knew I would be on my back for at least 5 weeks and there was nothing my body could do to hasten the healing. What I complained about I will describe in a later chapter. What was eating away at me was the impersonal nature of the contacts I had with the personnel. I was told that a geriatrician spent two half days a week in the nursing home. She never came to see me except for the two times I demanded to see her. And when I told her line and verse how poorly selected and trained the aides were, she agreed with me!

I will now say something readers may find hard to believe. Almost never did any aide or nurse initiate a conversation with me. As always, there were exceptions, in this instance two aides: Wayne Howard and Dana Kendrick. They are remarkably similar people. They are both very bright, have a realistic positive opinion of their worth, are sparkling, spontaneous, and curious about who you are and what you did in life, and have no hesitation responding to your questions about their background and future plans. And that is how I learned that my assessment of the nursing home was on target. On each day and for each of the three shifts, the nurse in charge forms teams, of usually three aides, and assigns 10 or so patients for whom they are responsible. It is rare that you will have the same team today that you had yesterday or you will have tomorrow. So when Dana Kendrick or Wayne Howard came into my room, I thanked God for big favors.

In the phenomenology of the staffs of nursing homes, their prime obligation is what they can do to repair bodily conditions which cause pain and suffering. And "do" refers to medications, diet, transfusions, blood tests, and frequent monitoring of vital health signs. What is rare is another "do" factor: What can I do to develop a relationship with the patient? What can I do to determine whether he or she has thoughts, feelings, fears, and anxieties which can interfere with recovery? Does their withdrawal and negative affect reflect an underlying condition or mirror my approach to their treatment and care?

In the medical community that question engenders guffaws, or anger, or the accusation that your heart is in the right place, but your mind is up in the clouds. We are not psychotherapists. We are very busy people, and our one and only goal is to get our patients well as soon as possible. That is what they expect and that is what they get.

I will now give my standard reply. The names of doctors I refer to are their real names.

I am not suggesting that you act or become a psychotherapist. What I am suggesting is that the quality of the interpersonal relationship between doctor and patient often plays a role—sometimes minor, sometimes major—in a patient's coping with his or her illness. The doctor may not recognize it but most patients do. For example, why is it that so many of Saul Milles' patients are doctors? Saul is an excellent internist-cardiologist. New Haven has a lot of such specialists so why is he known as a doctor's doctor? I have been a patient of his from the time he began a practice. You can say he is a good guy, but that is too vague and unrevealing. I would describe him as mild mannered, relaxed, the opposite of pontifical or God's gift to the healing profession, who does not interrupt you when you try to describe your symptoms, he does not talk down to you from an Olympian perch, his body language makes you feel that he has all the time in the world to be with you, he has a sense of humor, his eyes never leave you and he is listening with sincere interest in all you say and not with any indication of impatience. You quickly feel that he is someone you can trust with your feelings and fears.

And what about Dr. Leonard Farber (oncologist) and Dr. Nina Horowitz (surgeon)? Nina is the opposite of the stereotype of surgeons—who are interested only in the innards of your body and your feelings and fears are "noise." And is not Leonard Farber a warm, friendly, and sympathetic person whom his patients adore because he was there for them when they needed him the most?

Can you deny that the type of relationship I am trying to describe has no consequences at all for the recovery process, that body and mind are on two separate planets? In the abstract, you know better but the abstraction remains an abstraction.

What I related above is why a decade before I retired I wrote a book, *Caring and Compassion in Clinical Practice* (1986). It was about physicians, clinical psychologists, clinical social workers, and school teachers. School teachers are not considered clinicians, but in the decades I spent in school classrooms, I learned otherwise. That is not to criticize teachers because to do so would be an instance of blaming the victim and that is another story I cannot go into in this book (Sarason, 1992).

There are people in the community who are not clinicians or professionals, but who in their network of friends are seen and used as someone to whom they can bring personal problems and know that that person will try to be helpful and observe the obligations of confidentiality. In a paper published in 1978, Dr. Alan Towbin, a clinical psychologist (and a dear friend), ingeniously located and interviewed a sample of what he called "confidants."[3] What were the attributes of these confidants? They were precisely what I was trying to say to uncomprehending physicians.

If the physicians had asked me what credible evidence I could show them to support my argument that the quality of the interpersonal relationship played a role in healing and recovery from illness, I would have little credible data that would stand up in a court of evidence, and I would have been unable to do so. Physicians have long derogated my line of argument and have pointed with pride at the success of medical science to reduce the interfering role of psychological factors in treating patients. And you might ask them how come, with all that success, patients still see their interpersonal relationships with physicians as unsatisfactory for one or another reason. For example, when I entered the nursing home, my orthopedist, the geriatrician, the nurses, and everybody else predicted I would be in the nursing home 12–18 weeks. So when I left the nursing home after "only" 8 weeks, everybody proclaimed I was a success story. I saw my stay there as purgatory. It was not a case of the operation being a success but the patient died, but it came perilously close to that. Were it not for some dear friends, I would have sought means to end it all. I trust this will become clear when I describe my 2 months in a densely populated place where impersonality reigns supreme.

---

[3]Dr. Towbin's paper was published thirty years before that of Andres-Hyman, et al. I find it strange that given the title of Towbin's paper and the journal in which it appeared, his paper is not referred to.

So let me end this chapter by calling the reader's attention to a paper, the most recent to date, of the issue I have been discussing (Andres-Hyman, Strauss, & Davidson, 2007). What makes this study so instructive and important is that it is not about physicians, but psychotherapists. From the time clinical psychology formally became a recognized field in graduate education, a debate began which was stimulated by the recognition that all of the major theories about how to bring about a successful therapeutic outcome presented findings to support their approach. Whether it was Freudian or Jungian or Adlerian therapy, or Rogers' non-directive therapy, or Skinnerian behavior modification, or Ellis' Rational Emotive Therapy, or diverse versions of group therapy—all of them could claim an encouraging rate of successful outcomes. Each of these approaches dictated the structure of the therapeutic relationship as well as the behavior of the therapist and the techniques he or she would employ. Put it this way: Each approach had a "manual" describing what the therapist should do to repair something gone astray in a person's psyche. The manual is only a first step which is then followed by supervision of your work by a senior therapist who will point out to you where and why your understanding of the manual is superficial, a distortion, or just plain wrong. The unlearning and learning process can take a long time. (It is no different with auto mechanics, chefs, etc.)

If each approach has some success and no approach is a raging success, are there common factors in their manuals which are not recognized or are under-appreciated? Do those common factors throw light on why these manuals and training, each approach taking pride in their theoretical–technical uniqueness, are missing the forest for the trees? These are the questions Andres-Hyman et al. endeavored to answer by reviewing a large research literature on therapeutic outcomes. They very succinctly state their conclusions in these opening sentences of their paper.

> Research underscores the central role of factors in healing that appear to relate to the therapeutic relationship. These nonspecific or common factors and placebo effects are often overshadowed by an emphasis in the field on particular empirically supported therapies. Yet relationship variables account for a greater proportion of the variance in treatment outcomes than the technical intervention employed, representing a notable blind spot in our science and, by extension, our practice. As a consequence, clinical instruction in psychology and in the health professions more broadly generally lacks adequate specificity with respect to how to cultivate a healing relationship.

What you are as a person, how you come across to clients, how wittingly or unwittingly (naturally) you say or do things that convey that you can empathize with and have respect for the other person's feelings and plight, that your suggestions and advice are matters to be pondered and discussed and not as diagnostic utterances from someone who has cornered the market of truths, and you are, wonder of wonders, capable of admitting that you might be wrong; these are some of the common factors in the development of any close relationship, therapeutic or otherwise. It was the role of these common factors that enraged me in my 2-month stay in the nursing home. Having said that, I hasten to say that throughout that stay I knew that I could not blame anyone who worked there. Knowing as I did and do how people in these facilities are selected and trained, and the stereotypes about "Old Crocks" that abound in the society, I tried hard not to indulge in the blame game, although I was not always successful.

I have reason to anticipate that what I will be describing and concluding are unjustifiable overgeneralizations from personal experience. It may surprise the reader that I initially gave serious thought to writing a long chapter to show the comparison between the inadequacies and low quality of care in nursing homes with the failures of reform efforts to improve educational outcomes of our schools whose self-proclaimed mission is to help every child develop his or her potential. Beginning in 1965, orally and in print, I predicted with 100 percent accuracy why reform efforts would fail. My argument fell on deaf ears. At the risk of being too succinct, here is what schools and nursing homes have in common.

1. In our schools, students spend all of their time with teachers who are products of preparatory programs which ill prepared them for the realities of the culture of the school; in fact, few (if any) teachers have ever denied it. In their hierarchy, teachers are at the bottom of the pyramid of status and influence. More than half of all new teachers leave the field after 5 years.

2. For all practical purposes nursing home aides are neither selected nor trained; they are undereducated, have few alternatives for employment, and are regarded by nurses and administrators as undependable and as harmful or neglectful. Their rate of pay is low. They have no power or influence at all in almost every aspect of their work. The turnover rate for aides is very high.

3. Up until 1960, the salaries of teachers were scandalously low but started to steadily increase and today are more than barely respectable. That increase was explicitly justified by the argument that it would bring about an improvement in the student learning and acquisition of cognitive skills. Those expectations have not been manifested. Early in my stance was that the increase was justified on the grounds of comparative equity and would and could not alter the quality of learning or outcomes. I have made the same predictions in the case of aides.

4. For aides, nurses, and teachers, there is one factor which is a colossal barrier to the goals of their mission statements: Time. Teachers are given a curriculum which they must cover in a prescribed point of time and most teachers are fearful that if the students do not do well on state-mandated achievement tests, it may have very negative consequences for the teachers and the school. The No Child Left Behind reforms spell out in clear detail what the "shape up and ship out" consequences will be. Teachers spend their time teaching to the test, drilling students as if the classroom was an army boot camp. Teachers have no time to work with individual students who need special attention; they know and feel guilty that they are colluding in a charade in which what is all important is that students pass all the important tests. In nursing homes, nurses spend hours filling out forms required by the state health department and going to each room to give patients their medications (making the conventional imagery of nurses as caretakers hardly in evidence, when a patient sends a signal to the nurses station for one or another reason, the nurse dispatches an aide to the patient's room: From the standpoint of the patients, all relationships are fleeting and impersonal.

Nurses are dispatchers. Orders have a "curriculum." They are given a list of patients—it is not a short list—whose needs must be met before the next shift of aides appear. It is a rare day that an aide does not have a patient that requires an "unusual" amount of time which means that her other patients will not get the 'usual" amount of time. From the standpoint of aides, the needs of patients are narrowly physical, not social or psychological.

I decided against writing a comprehensive chapter comparing nursing homes with schools and other human service institutions because it might deflect attention away from the central concern of this book. Besides, the history of nursing homes is a grim one about which the powers that be are either ignorant or deluding themselves by the belief that the future will take care of itself, a delusion in the category of famous last words. It took half a century for global warming to begin to be taken seriously. That the average quality of nursing homes is poor, that they are crowded places, that the poorest ones cannot be closed because there are not available beds elsewhere; that millions of baby boomers are at or near old age and many will need or want a protective environment containing independent and assisted living apartments, a nursing home; that a significant number of middle-class baby boomers will be unable to afford the entry fee or monthly service charge these total care facilities require—in the halls of political power and responsibility and planning, not a single word has been uttered about any of these brute facts. And they are facts, not the inventions of a doom and gloom personality. It was about those facts that is center stage in this book.

As a psychologist, it would be professionally suicidal to deny that I have not cornered the market on truth, objectivity, and wisdom. Yes, my 2 month's stay in the nursing home in the total care facility in which I reside was traumatic, but not because the nurses, aides, and higher echelon administrators were villains. No, it was because their preparation for their responsibilities for aged people rendered them incapable of comprehending the phenomenology of aged, dependent, fearful people who know their days are very numbered, who yearn for and seek mutually supportive relationships which diminish the boundaries of an unwanted privacy, who know the differences between intimacy and ritualistic small talk. They need and want what people of all ages need and want, but in the aged person, the strength of their needs and wants is exponentially and poignantly stronger as their feelings of self-worth decline. Caretakers are unaware of all this. They are taught to see aged people through prisms which direct their attention to physical symptoms and decline. They regard the obstreperous, demanding patient as undeserved punishment from Divine Providence. They are grateful to the same heavenly source for the silent, passive, unresponsive patient. The two types of patients represent extremes between which are patients who for the most part have been able to take seriously what one patient said to me, "There is no point in fighting city hall. It has no payoff."

In light of what I have said in this introductory chapter, I feel compelled to emphasize that I regard nurses and aides as victims and not as causes of what I will describe in later pages. They are victims of a selection and training process which is a mockery of what is meant by the concepts of selection and training. No less important is the organization and culture of nursing homes regardless of whether they are or are not part of a total care facility. All nursing homes are creations of and monitored by the state department of health, which determines the criteria for selection and training as well as the criteria for quality care of patients. It may come as a surprise to the reader when I say that what I describe in this book is old news to the personnel in the state department of health. For reasons I discuss in later pages, there is little they can do to diminish the disconnect between what should be and is. They know full well that the problem is not nursing home personnel but a *system* (which includes legislatures and governors), the parts of which are in an adversarial relationship with each other. As in the case of our schools, the intractability of nursing homes to reform efforts is incomprehensible unless you take seriously the adversarial nature of the system in which they are embedded.

Writers of nonfiction books rarely end their first chapter by urging their readers to read a related book. I have chosen to do so because it is the only book I know which is about nursing homes that was written by an aide! It is a book written by Thomas Gass (2004) with the title *Nobody's Home. Candid Reflections of a Nursing Home Aide*. It truly is a very well written, poignant, and even inspiring book. Fortunately, for my purposes Gass' focus is exclusively on the daily routine, events, happenings, and relationships. It is a literary portrait of Ground Zero painted with feeling, sensitivity, understanding, fairness that does not obscure the wide range of human frailties. It is the opposite of a polemic. As you read the book, it is hard to avoid the feeling that Gass has to exert control over feelings of outrage. He rarely, if ever, plays the blame game.

The nursing home in which I spent 2 months is part of a total care facility containing apartments for those capable of independent living. The residents are highly educated people, almost all of whom belong to the haves of this world. The nursing home in which Gass was an aide is independent, not a part, of a total care facility; the patients were certainly not highly educated and from what Gass describes they are among the have-nots of this world and most of them are psychologically fragile, some with psychotic symptoms. Gass tells us little about their previous lifestyle.

Comparing the two nursing homes is like comparing a local storm and a hurricane. When Gass describes his "ordinary" duties as well as the problems of patient behaviors he had to deal with, he never acts angry or expresses outrage. He sees his responsibility to relate to patients in ways that will dilute their low sense of personal worth and self-respect. What his book conjured up in me was the imagery of a saintly man of action. It is as though he truly believes that he is called by God to do God's work. He knows—how could he not know?—that he is very different from all other aides, but he never directs criticism to or about them.

When you read my chapter about my stay in the nursing home, it will be markedly different than what Gass describes. Some of the reasons are obvious. One reason is that, in mine, approximately 70 percent of the residents (men and women) have had very productive lives. Their level of self-respect and worth is high, they are not shrinking violets, and they do not fear expressing their dissatisfactions. They rightly feel that they are not paying a lot of money to keep their mouths shut when they have a grievance. And when they have grievances which have not been remedied by lower echelon of power, they do not hesitate to go to the CEO of the Center. And the grievances are not few, some minor and tolerable, some as I shall describe major and festering. No resident in the Center came to it with the expectation that they were entering a heaven on earth. There was one thing they knew in an abstract way: Any organization, especially those we label as human services institutions, is bureaucratic and stratified, power flows from the top of the organizational pyramid in decreasing amounts to each of the lower strata. The larger the organization the less the top strata knew about the lower strata. In a later chapter, I discuss what I call organizational craziness. Institutions vary in the level of craziness, but there is no human service institution anywhere near zero level, and the grievances of residents almost always reflect the craziness. That is a major point I discuss in the early chapters of this book.

In the early years of my career, I worked and lived in human services institutions where I got to know aides. In subsequent years, I was a consultant to a variety of such institutions. I came to know well scores of aides. I have not met Tom Gass. But I could not avoid concluding after reading his book that he is unique among all aides I have ever known. So let me tell you what Gass tells us about himself.

> I am not a conventional nursing-home aide. I am not conventional in most ways. I spent five years in a Catholic seminary, which included a year of silence. Afterward I went to university and graduated with a psychology degree. I taught on an Indian reservation and served as director of a halfway house. I spent seven years in a meditation community. My interest in Eastern philosophies led me to spend an itinerant decade in Asia. When I learned my mother was dying I returned to America to care for her. The experience of attending her death process affected me. I felt that I had to perform some kind of meaningful work.
>
> I entered this work with the vague idea that it would make me useful in some way that I had not yet discovered. When I hired on, I boldly stated that my motivation was spiritual: "I'm not particularly interested in being cheerful or sweet or polite. I just ask for permission to be real." I came hoping that I would gain some depth of character by doing good.
>
> My sense of service came not purely out of altruism but from a constant fear of being not good enough. I was raised by family and religion to believe there was an essential defect in my nature. So I acted as if, by bowing deeply, in humble service, thus giving myself away, I might somehow purchase my own goodness.

That explains why in the following chapters there are two major, interrelated foci. The first is that there can be no improvement in quality of the care of aged people unless there is a radical overhaul of the ways we select and train their caretakers: aides, nurses, administrators, and physicians who are geriatricians. The second is that it is a statistical certainly that millions of baby boomers will soon be senior citizens and need a protective environment, which today is in short supply, and much of which, to be charitable, I will say is unacceptable. A very knowledgeable colleague who read the manuscript of this suggested that the subtitle of this book should be the coming disaster in the care of aged people. That sounded too apocalyptic to my ears. Given my advanced age, it is a statistical certainty that I will not be around to know if it is a crisis or a disaster. Either way, the outcome will not be a pleasant one.

The early chapters of this book are very personal and may lead some readers to conclude that I am exaggerating the overall inadequacies of nursing home care. In later chapters, I present data to confirm my conclusions. But those data tell only part of the story. Let me very briefly explain why.

1. Nursing homes are regulated and monitored by the state department of health.
2. Regulations are reflected in standards (often in numerical terms) intended to safeguard the physical health of patients. The standards literally define "humane care." The standards say next to nothing about the substance and quality of interpersonal, social, and psychological issues inherent in the care of aged people.
3. The state department personnel are very much aware that they have insufficient staff for the monitoring function. The larger the state, the more monitoring function is a mockery, a charade.
4. The state department is empowered to fine, put on notice, and close a nursing home. In Connecticut, the state department prepares a list of those nursing homes which have been penalized for infractions. Any citizen who requests the list must be given it. The state department is not required to make the list known to the general public (unlike what the state department of health is required to do for all hospitals in regard to deaths following certain surgeries).
5. It is by no means unusual for a nursing home in danger of large penalties for multiple infractions to seek and obtain the support of local and state politicians to alter the penalties or to markedly extend the amount of time the home will have to make changes to a process.
6. The state department is between a rock and a hard place. Its staff is not in doubt that there are nursing homes that should be closed, but in light of the shortage of beds, the patients would have to be scattered all over the state, a process tantamount to inflicting insult to injury. Many will not survive.

Aged people have minds, needs, feelings, desires, and a very acute sensitivity to the fact that their days are very numbered. They yearn for understanding, friendship, intimacy, respect, and understanding from their final caretakers. They know the bells will toll for them in the near future. They know all that. What is piercing and poignant to them is that in the nursing home their caretakers do not know "all that," that caretakers care about bodily afflictions, blood tests, medications, meals, and changing the bed sheets, all of which is done as quickly as possible because time is of the essence and what you feel and think about is unimportant "noise" that disrupts a routine. To the caretakers of aged people, it is foreign territory that remains unexplored and they have no need, desire, or curiosity to explore. Aged people and their caretakers are like strangers passing each other in the night.

I am not blaming the caretakers. Over a career of six decades, I was in a teaching relationship to physicians, nurses, and aides in diverse human service settings, including hospitals, nursing and medical schools. I know how these caretakers are selected and trained in a culture in which the psychological and interpersonal are, if not derogated, given short shrift. In recent decades, the rhetoric has changed, but that is an instance of the adage that the more things change the more they remain the same. That is why in this book, I emphasize that unless and until the selection and training of the caretakers of aged people are radically changed, nursing homes will remain impersonal warehouses. I try to avoid thinking what will happen when from the millions of baby boomers who in a decade or two will become senior citizens, there will be a sizeable number who also require a protective environment. Such a dramatic increase is, demographically speaking, a statistical certainty. You can bet the ranch on it. So let me close this chapter with a study I did of every national election beginning with that of 2000 and ending with that of 2006. I heard and saw on TV approximately 70 percent of the speeches candidates gave. I was interested in only one question: How many times did candidates say anything to indicate that (a) the quality of care in existing nursing homes was nothing to be proud of, (b) in the next decade or two the problems will get worse as the baby-boom generations will be well in their "golden years."

What were the results of my "study" of the speeches of candidates for congress and the presidency? The numerical number is zero! Yes, they all proclaimed the mantra that senior citizens must be assured that their Medicare and social security benefits will be protected. The mantra belongs in the category of famous last words. I hope that after reading this book, the reader will understand why I wrote this book.

Chapter 2
Becoming a Resident in a Total Care Facility

This book raises critical, troublesome questions about how policy makers are thinking about and planning for the fact that in a decade or so there will be a cohort of aged people numbering not in the thousands but in the millions, a trend that will continue for several decades. It inevitably will transform American society. That transformation will be no less dramatic than the waves of immigrations in the nineteenth and early twentieth century or the less-studied and -discussed consequences of the GI Bill for the millions of citizens who were in the armed services in World War II.

Today we are witness to a transformation powered by the millions of immigrants who live here, legally or illegally, and were born in Mexico and Central America, a fact that is causing intense debate in Congress and the country. That influx has already changed the political scene in many states, especially in California and in the southwest states. The debate has many facets: political, economic, educational, cultural, religious, constitutional, historical.

For the past half century, demographers warned us that by the end of the twentieth century the millions comprising the baby-boom cohorts would begin to approach retirement and look forward to enjoying the "glories of the golden years." At the same time, economists were asserting that around 2015 the social security fund would be paying out more than it was receiving. Then came 9/11 with the result that, instead of predicted budget surpluses, the government began to incur steep deficits which mightily increased an already huge national debt, in part because of the president's earlier steep tax-reduction program. Millions of the baby-boom cohorts look to an imperiled economic future. In the course of their lives, Americans save less than those of any other developed country. Many members of Congress as well as many economists advocate cuts in spending for social programs and that includes, of course, programs for senior citizens.

It is playing loose with words and concepts to say that the government has a program for those who want to or will be required to stop working. As noted earlier, at age 85, my physical condition led me to decide to enter a total care facility where I could continue to live independently, but which also had a nursing home, in the event that my health deteriorated. I entered such a center at the age of 85. At age 88, I fell, fractured my hip and pelvis, and spent 2 months in the nursing home before I returned to independent living in my apartment. I observed and experienced a lot of

S.B. Sarason, *Centers for Ending*, Caregiving: Research, Practice, Policy,
DOI 10.1007/978-1-4419-5725-2_2, © Springer Science+Business Media, LLC 2011

the nursing home, enough to conclude that there were serious problems and issues about which policy makers seemed totally unaware or unable to do anything. By policy makers, I am referring to the politicians who enact these programs, the state agencies which supervise facilities that are created, and the professionals and staff who work in these facilities. I know there are exceptions to this criticism, but I can assure the readers they are just that: exceptions.

The issues and problems to which I refer cannot be adequately conveyed by resort to generalizations. So I decided to begin this book with concrete descriptions of my 2 months in a total care facility which has, as almost all such facilities have, apartments for whose capable of independent living, single rooms for those who with assistance can make do, and a nursing home in a separate building.[1]

As centers go, the nursing home in which I was resident had about 50 beds. The assisted living unit had about 20 residents. Those capable of independent living number about 200 living alone or with a spouse. Compared to other total care centers, the one I chose is below average in physical size and number of residents.

The reader may rightly ask: "What permits you, a single individual, to use your experience as applicable to such centers generally?" There are two answers to that question. The first is that over a long career, I have worked in and consulted to a variety of total care facilities serving mentally handicapped children, and I consulted and got to know well an institution for juvenile delinquents. And 30 years ago, I had to bring my father to New Haven to place him in a nursing home, which I got to know extremely well. The one I chose came after I had visited a number of nursing homes.

The second answer is that during and after my 2-month stay in the center's nursing home—sharing experience with other residents—and later meeting knowledgeable people in neighboring states, I was confirmed in my opinion that total care facilities should be considered guilty until they can demonstrate their innocence.

---

[1] When I was admitted to the center in which I reside, I had to pay an entry fee of $166,000, none of which is refundable after the first year. The monthly maintenance charge, which has been increased each year I have been here, is currently $3,600 dollars. My income from my pension and social security is $53,000 per year. When my closest Yale colleague and his wife chose to live in a new, larger and more posh center, they had to pay up front $450,000 dollars, a percentage of which will be paid to their heirs when he and his wife die. Their monthly maintenance charge is $4,500 per month. Of all centers in Connecticut mine is average for what it costs and my friend's center is near the top. Mine is a nonprofit one; his is for profit. The legendary Senator Everett Dirksen said, "A million here and a million there and before you know it you are talking about money." Although states and regions vary in average costs of living in a total care facility, the variations would be less if you did not include nonprofit ones, such as those sponsored by religious groups or universities. Even if you do not take account of inflation, a recession, or (God forbid) a depression, you do not have to be an economist to conclude that a significant fraction of people who consider themselves middle class will be unable to afford living in a total care facility. Despite what I will say and describe in later pages I feel fortunate to be living where I am, especially when I think of the alternatives I might have been forced to consider. But being fortunate is not to be confused with being content or happy. On November 10, 2006, Hedrick Smith presented a one hour Frontline documentary (PBS Org.) on the inadequacy of 401 K retirement plans which many baby boomers who consider themselves middle class contribute to. They are in for a shock.

I will have more to say about this in a later chapter. The history of these facilities in the post World War II decades is the opposite of fun reading.

What is emerging in the arena of services and facilities for aged people is two tiered and the gulf between them is stark. One tier is comprised of people who can afford the costs of living in the total care centers which have been built and are being built at an ever increasing rate. Poor people or those of modest means cannot afford to live in them. What happens to those people? I take this up in later pages.

Entering the Center

Why did I decide to move to a total care facility? I had been living in a lovely condo development consisting of 450 ranch type houses each of which had two spacious units and a finished basement. I moved there in 1995 because I had a friend there. Prior to that, I lived with my wife for 50 years in a New Haven suburb. I retired from Yale in 1989. On June 7, 1993, we were in an auto accident: my wife was killed instantly and I spent 3 months in three hospitals. When I returned to my home, I tried to adapt to the ever present feeling of loss and utter loneliness. My daughter and her family lived in Lowell, MA. My brother lived in Seattle. I had many friends, but it in no way lessened the feelings of loss and loneliness. There were times I wanted to end it all. What kept me going was the activity from which I have always enjoyed the greatest satisfaction: writing. The condo development seemed ideal for me and I moved there. My friend, who is a very busy person, and I saw each other on weekends, but the bulk of the time I lived alone in my very comfortable condo with no one to eat with, talk with, or sleep with. I made friends with some lovely people. I could be alone and tolerate it, but this sense of loneliness always intruded. I did realize that I was fortunate to be where I was, but my need for intimacy never lessened.

Then all hell broke loose with my eyes. Macular degeneration took over; I lost vision in one eye and central vision in the other. I could not drive. My visual world dimmed. Writing became difficult, I could not read, and several times, I fell because of the impaired depth perception caused by macular degeneration. I bit the bullet and decided I needed to be in a total care facility. Someone once said that you move to a total care facility when you are still able to walk in, not when they have to carry you in.

It is extraordinarily difficult for people to talk about the need for intimacy. It is as if it is an admission of weakness, a character trait they do not want to reveal to others. In the years when I was in a psychotherapeutic relation with clients, these feelings were very frequently reported. The novel *August* by Judith Rossner is about a woman whose therapist took summer vacations in August and the difficulty she experienced with her loneliness and need for intimacy.

It used to be that marriage took place between men and women who lived in the same city or neighborhood or via family networks or individual "match makers." This was especially true of the sons and daughters of immigrants.

There were times when getting a divorce was a difficult, long, and costly process as well as frowned upon because of religious reasons. World War II changed all that. During the war years, there was a shortage of men for women to marry. And the millions of men who were in the armed services saw the world of women and marriage radically different from when they were young civilians. In World War I, it was put this way: "How can you keep them down on the farm once they have seen Paree?" World War II was followed by a seismic social change, one aspect of which was a dramatic alteration in how men and women viewed and experienced intimacy, sexual relationships, and marriage itself. Within 5 years after the war, the divorce rate began to increase each year. Interpersonal relationships between the sexes became more shallow and short lived at the same time feelings of loneliness and the absence of intimacy were exacerbated. Men were reluctant to be "imprisoned" in marriage and women were reluctant to be married before they had begun a working career.

The above digression to recent history was prologue to a question put to me by single and/or divorced friends, graduate students, and young faculty members. That question was: "How can I meet people whom I might find interesting and I might want to marry? Where are you likely to meet them? I don't want to live alone. The thought that I might have to resign myself to that kind of existence drives me up a wall. The kind of mate I want is usually already married. The kind of person I have dated is not interested in marriage but in an affair which I do not need."

That was a question I began to be asked around 1970. I was no longer doing clinical work, but I had many friends who are clinical psychologists, psychiatrists, or marriage counselors. The quick answer I got from all of those colleagues was very succinct: "I have many such patients and if it wasn't for them I would have to give up my practice and go on public welfare."

I assume that readers have heard on radio and TV commercials for companies that have developed measures that allow them to match a man and a woman "who were meant for each other." The commercial says nothing about how the company will arrange a meeting with someone whose interests and personal characteristics are compatible to yours. Nothing is prominent except a happy life with the person of your dreams. Nothing is said about cost. Commercials on radio and TV do not come cheap. One has to assume that thousands of persons have responded to these commercials and that such companies do not write the bottom line of their tax return in red ink.

Over the course of a lifetime, every human being has an undetermined number of experiences at the core of which is the fear of loneliness and the absence of accustomed sources of interpersonal intimacy. Someone once said that if you want to place blame for the psychological problems of life in highly developed countries, you should place that blame on the person who invented the process whereby steam could be transmitted over a distance. That process severed the relationship between where one worked and lived. It changed the dynamics and structure of relations among family members. The mills and factories of the industrial revolution in which men, women, and young children worked many hours of the days, 6 days a week, dramatically changed the quality and intimacy of family life and caused a future perspective that was cloudy, unpredictable, and dangerous to mind and body.

It was the beginning of fault lines within the nuclear family as well as its boundaries with the extended families. Those changes became obvious when steam-powered engines made transportation to distant places possible. In Lowell, Massachusetts, the National Park Service has converted its old nineteenth century mills so that visitors can see slide shows showing how owners of mills lured young European girls, mostly from Ireland, to come to work in the mills. They lived a life of imposed loneliness and frustrated intimacy. It's not a pretty story. And that story should be viewed in relation to the waves of immigration that would soon bring millions of people to America who would experience rejection, loneliness, and face conditions not conducive to the need for intimacy. For most of them, their only source of intimacy was with their God.

We are descendants of those immigrants on the basis of more than blood. I cannot expand on that assertion here. Here again, novelists have described and more than touched on what I have said. Willa Cather, Edith Wharton, Jack London, Theodore Dreiser, Edward O'Conner, and the playwright Eugene O'Neil come to mind.

So what does all this have to do with my decision to live in a total care facility? Yes, I had felt more lonely and the absence of intimacy more than ever before, but I reluctantly adapted to such feelings. Indeed, I had concluded in my 1974 book *The Psychological Sense of Community* that what I called "unwanted privacy" plagued far too many people in contemporary society to a disturbing degree. The death of my wife in 1993 put me in that category. I had and have many dear friends, near and far, male and female, young and old. When I see them—usually once a week, often after a longer interval—and talk matters of mutual interest, I consider it inappropriate and unproductive to talk about what is missing in my life. They will listen, they will feel badly about what I would tell them, but they and I know they are not in a position to substitute for what is missing in my life. Catharsis is highly overrated.

I entered a total care facility for reasons of health, personal safety, and to live within very close proximity to other residents who had come for similar reasons. I would not be spending my days alone in my apartment. It was in marked contrast to having a residence in a Manhattan apartment house where it is not uncommon that you know practically no one in the building. When I told my plans to friends, some of them, all much younger than I, said, "Watch out. The large majority of the residents will be women who outlived their husbands. They will be after you. You will have to fight them off." Two friends who are women as old as I am said the same thing even though both say they have no desire whatsoever for anything resembling an intimate relationship. Catch an eligible man? That they would never, but never, do. However you define sexuality, these two smart, lovely women want little or nothing of it. In any event, I entered the center of endings with relief and curiosity.

The first decision I made was that for 5 months I would go to the dining room and sit down at a table that had an empty seat. I would introduce myself, tell them where I had been living and why I came to this center. I would also endeavor to get them to tell me where they were from and how long they had been at the center. The result was that the conversations were for me unrelieving; generally speaking, it was not an ice breaker except when one of them turned out to have been a faculty member or an employee at Yale whom I never had met.

Interpersonally, it was not an easy or satisfying 5 months. I felt somewhat like younger people plagued by feelings of loneliness and the absence of intimacy who ask: "Where can I go to meet someone with whom I can have some kind of satisfying relationship?" The center has no dating service! You live alone in your apartment and your social life is up to you. You can join the book club, or the poker, bridge, scrabble, stitch and knit groups. The center has an excellent library, but most of the time you will see no more than three people in it. There is a weekly lecture series, on Saturday night there is a movie chosen by a committee who frown upon movies containing obscene language and/or explicit sex. If you can drive, you can see movies in a cinema complex. Most residents do not have a car, most often because of poor sight. At lectures and the Saturday movie, there are always some residents who fall asleep, often because the lecturer or the film facilitates sleep. They are rarely stimulating.

There is a lot "to do" at the center, but it is hard for me to avoid the impression that people "do" these things to fill time of which they have a great deal. This is particularly true of those women who came to maturity before the women's liberation movement. They were homemakers, did not go beyond high school, their spouse died, their children live far away. They had developed no interest in the affairs of the world. They find others like them and they band together. There is a much larger group of highly educated women who had held very responsible positions.

There are about a dozen married couples at the center because the spouse has a serious health problem. Finally, there are about 20 single men who are either frail or are handicapped in some way.

A caveat is in order here. I did not enter the center to study residents and the center as a social institution. I was curious about that. But I was too preoccupied with my health problems, my remaining years, and the fear that my productive years were over. After my first week in the center it was obvious that it would be a full time job to try to meet the scores of people there, let alone establish a relationship with even a few. Because of my limited vision, associating names with faces was very frustrating, indeed distressing. There were individuals with whom I thought I shared common interests. I took the initiative to try to get to know them, but only one of them subsequently sought me out. There were people whom I liked and respected, but for reasons I could not fathom; I would interact with them only when we happened to sit at the same dinner table. Was my personality a major factor? All I can say is that over a long lifetime I have made enduring friendships with a wide variety of people. Making rewarding friendships in your waning years is dramatically more difficult and infrequent than in your previous years. I was fortunate that several weeks after I came to the center, a woman came over to the table where I was eating supper. She smiled and said that I probably did not remember her. With my eyes, she was correct, but when she told me her name I jumped with joy. I had known her more than 20 years ago and I admired her. Shortly after I entered, two people I had known and who had the same kind of interests in public schools as I did became residents at the center. What was also helpful (too weak a word) was that my next door neighbor whom I had not known well—she was a cousin of a dear friend— spontaneously and graciously became a trusted friend. Those four women made life in the center tolerable. They made my adjustment to the center less difficult. They were not just friends, but intimates.

What grounds do I have for generalizing from the 200 people I hardly know, some of whom I still have not met and who vary wildly in their overt appearance, personal style, health, background, and length of time they have been in the center? The one thing they know for certain is that they are old. I suppose we can bet the ranch that these facilities have far more women than men. And you can feel secure in assuming that all residents know they will die in the center. Stereotypes about old people abound and they are used by people (and even some medical professionals) to explain what old people think and do. Stereotypes are like clichés: They contain a kernel of truth that masks the variation among those whom the stereotypes or cliché purport to explain. Let me illustrate this with my experience with mentally retarded institutionalized people with whom I worked *and* lived.

I finished my graduate training in 1942 and took a position as psychologist in a brand new, innovative institution in rural Connecticut. I had read a fair amount about mental retardation; I had administered various intelligence tests to many such individuals. I knew how tests were used to place such people in certain groupings for purposes of program planning.

One-third of the resident population had from birth severe neurological, physical, and mental impairments, many were bed-ridden and were incapable of social-educational learning. One-third of the population was labeled high grade or moron, who on the surface looked normal, but whose cognitive ability was well below normal, at maturity would at best be capable of reading at a fourth grade level, and could be employed in the most routine and unchallenging jobs. They came from families whose parents and siblings were intellectually impoverished and family life disorganized. The prevailing professional view was that their cognitive impairments were inherited. In between these two populations were middle-grade or "imbecile" children: a hodge-podge of conditions which I shall not attempt to describe. What I shall now describe has to do only with the "morons."

What the reader must keep in mind is that I lived in the institution in close proximity to residents and professional staff. I ate frequently in the cottages where the residents were housed. I did more than administer tests. A resident who was acting up or strangely was sent to me to work my magic. If during my visits to classrooms in the school building or in the cottages, I saw a resident whose behavior puzzled me, I asked him or her if they wanted an appointment to talk to me. I have written of my experiences at the Southbury Training School in two books (Sarason, 1949, 1988). I also want to refer the reader to a book (1948) *The Unfolding of Artistic Activity*. It contains a long chapter on the artistic productions of a group of Southbury residents. That book was by Henry Schaefer-Simmern, a political refugee from Nazi Germany who worked over 2 years with his small group of residents. The book is a myth shattering, compelling one. One of the last things John Dewey wrote was an introduction to the book. I learned more from Schaefer than I did from my graduate training.

To put it succinctly, I had to unlearn everything I had read about mentally retarded individuals. I stopped seeing "moron" through the prisms of tests. I listened and learned about how and why they did not want to be in the institution, their longing to be with family. They were astute observers and critics of the inadequacies and unfairness of their teachers and supervisors. They yearned for intimacy, they sought sexual partners. Someone once said that prisons cure only one thing: heterosexuality. That was happening at the Southbury Training School where any overt expression of sexuality was forbidden and punished.

Not for one moment was I alerted in my training to the predictable problematics of institutional living where those who live and work there differ on a whole host of psychological, social, and stylistic variables and where, as in all institutions, there are different strata of power and responsibility. In human services institutions those who are recipients of care have no power. At the Southbury Training School, the residents had no power at all. Their daily lives were regulated and supervised 24 hours a day as long as they reside there, which can be years. Those in the upper stratas of power and/or responsibility have no doubt whatever that they "know" their residents as individuals and also know what is best for them. It is a total care facility, with the emphasis on total. As I learned, they do not know their residents as thinking, feeling, individuals who daydream, fantasize, have memories of family life before coming to Southbury, and whose curiosity about the outside world is boundless. All this is unknown to staff.

The total care facility in which I now live is not, of course, like the one I described for mentally retarded individuals. Residents exercised independence of choice to come to reside here. They can partake of all activities at the center and if they have a car they can go where they want to go for as long as they wish. If they do not have a car, they can be taken by van provided by the center to attend the production of live drama in three theaters. Those vans also take them to shopping malls and supermarkets, churches, concerts, and to offices of their doctors. The center does a praiseworthy job taking advantage of special events in the metropolitan New Haven community. The center is 10 minutes by car or van to the Yale campus.

I have never known a visitor to the center who was not mightily impressed with its appearance, ambiance, and facilities (which include a pool and an exercise room). I have never known a resident who disagreed with them. And the administrators take pride at the level of satisfaction of its residents: they imply that they know, respect, and cater to them. You can put their stance this way: "We know you and what you need and want and you can trust us to do what is best for you." Succinctly put: "Papa knows best."

Let me illustrate this point by recent events in the center where I reside. I need first to describe a meeting I had with the director months before recent events.

You are entitled to one meal a day. It can be lunch or dinner; both of which from a menu of many choices and you can eat as much as you want. This and no other comparable facility would rate one star from Michelin. But the food is more than acceptable. It is good. No gripes on that score. If you do not eat that daily meal, you could, if you wish, eat lunch and dinner on another day that month. Rarely do people who miss a meal eat lunch and dinner to "use up" the meal they missed. Most residents prefer to eat lightly at breakfast and at lunch. (Each apartment in the independent living unit has a small kitchen, stove, and refrigerator.)

I had friends at the center over the years who invited me to have dinner with them. They would refuse my offer to pay. They would say that they had not eaten every day at the center and therefore he or she could use the unused day (or days) for invited guests like me.

When I became a resident, I learned that the policy had been changed: Guests could no longer be unpaying guests, the cost would be charged to the resident. And I also soon learned that many residents were unhappy about it and had reduced the number of guests he or she invited. Some of the residents had family who came from afar to visit and they paid for their meals. I thought the change in policy was unfair. If you could not use the allotted number of meals during a month, why should you not have the privilege to use them for guests and family? I could afford it but simply felt it was unfair.

I asked one of the administrators about why the policy had been changed. She replied that there were residents, mostly women, who would deliberately skip a meal in order not to have to pay for an invited guest. "It is not good for their health to skip meals and keeping them in good health is our priority." That explanation did not have for me the ring of truth. At a one-on-one meeting with the executive director I asked why the policy was changed and did he know some residents resented it. His answer was quick and decisive. "We were running a deficit and that was one of the ways we could reduce it." He then looked at me coyly and said, "I know how much money each resident has. As you know, each resident fills out a form in the application detailing their financial resources. I can assure you they could afford the change in policy." End of conversation. He did not say he was sorry if residents resented the change and he certainly did not know that one of the most plaguing fears of many old people is that something will happen to them or to the national economy which will dilute or wipe out their income from diverse sources. There were residents who grew up in the Great Depression and experienced or witnessed the shattering of lives.

The above was prologue to a surprise event at the center. Each month there is a "community meeting" at which the director reports on issues and plans of interest and importance to the residents. Residents are encouraged to ask questions and make comments. The meetings are usually not interesting or controversial.

At one of these meetings, the director dropped a bomb which would enlarge and transform the center.

Here in brief are the guts of what he said:

1. In order to stay competitive with existing comparable centers and those certain to be built in the near future, he had to come up with new plans. Over the past 2 years, he and the Advisory Board hired and met with an architectural firm in another state. He passed around architectural drawings of what the center would look like at the end of the 4-year project.

2. The main stimulus for the planning was that the nursing home unit was woefully inadequate and inefficient. The rooms were too small and too few, there was inadequate space, the shower room was desolate and unpleasing to the eye. The nursing home was unrescuably inefficient and would have to be demolished and put elsewhere. The new nursing home would be placed on a larger piece of land. It would be a costly project.

3. To help defray the cost, two new buildings would be added to the main six-story building. The nursing home would be built near one of the new buildings. The distance from one new building to the main building to the other new building to the new nursing home was not small.

4. The two new buildings would contain luxury apartments. He never defined what he meant by luxury and how the new apartments would differ from the existing two bedroom units.

The residents listened in stunned silence. For the next month or more the residents vented their criticisms, anger, and fears to each other. These are the contents of their concerns.

1. They had no voice during the long period of planning. What was the universe of alternatives the director and advisory board had considered?
2. No data whatever about projected costs were given. Every year the monthly charge has been increased. Was it not likely that the plan would require a higher rate of increase in the monthly charge which for some residents would be difficult to pay?
3. The plan seemed based on the assumption that there would be no serious down-turns in the national economy, an assumption that has no basis in the nation's history. What if a downturn meant that the number of people who could afford luxury units would shrink? Worse, would the center be in danger of going bankrupt? That would literally be a catastrophe for the residents.
4. At present the center is full and has a waiting list of clients for a two bedroom apartment which in number is now much fewer than one bedroom apartments. Since residents are not privy to a detailed budget of the center, how likely is it that the center is currently not running a deficit?
5. To help pay for a new Health Center (= nursing home), its units will be signif-icantly more expensive than any two bedroom apartment in the center as it now is. That will have two consequences. One possibility is that there will be two "classes" of people. It may not be a have vs. have-not dichotomy, but that will be how the present and future populations in the center may see it: a division along socioeconomic lines. The second consequence stems from the fact that the center is a non-profit enterprise with an advisory board (= Board of Trustees) the members of which live in or near New Haven. How can such an organization justify a plan that would make it impossible for people who need and can only afford a one-bedroom apartment?
6. The center's dining room is a very attractive place but it has all it can do to accommodate the current number of residents. Will each of the new buildings have their own dining facilities with whatever you call a more "upscale" menu?
7. The residents are in total agreement that we need a new nursing home. All of them know some residents will spend varying amounts of time in the health center for various reasons before they can return to independent living in their apartment. They may have an illness which requires hospitalization after which they go to the nursing home for treatment and rehabilitation. Injuries due to falls, surgery to deal with a bodily dysfunction, heart attacks, and strokes: these and more conditions require hospitalization followed by a stay in the nursing home. The residents now capable of independent living know they are not likely to be in their apartment when they die. All this explains why the residents reacted so strongly to where a future nursing home would be located. They pointed out that many of them tire easily, use walkers or a wheelchair. As it is now they are still able to visit their friends who are in the nursing home. But where the new nursing home was to be built would be extremely tiring or impossible. Dying alone and

away from friends was something to be avoided and one of the reasons they came to the center.

8. The center is located in an attractive wooded area behind which is a lake. Implementation of the proposed building plan is to be carried out over 4 years. As some residents pointed out, "For 4 years we will have trucks, cranes, dumpsters, and a lot of noise. Silence and quiet will become precious amenities. We did not come to the center to see the kind of mess this place will be for 4 years. We didn't put down a lot of money to be here, and a hefty monthly services charge to boot, to experience 4 years of aesthetically ugly turmoil." I cannot refrain from quoting what one of the residents said: "The administrators are now of an age when they very likely will be alive to see fulfillment of their dreams. I am old enough to say that at my age I may not be alive to dream."

There are several reasons I summarized a very complicated story. The first was that it illustrates how ignorant and/or insensitive the administration was about the phenomenology of people near the end of their lives, a fact confirmed by the controlled anger and puzzlement of the administration. It is more than a story of the top-down style of leadership and its consequences. It is about an age cohort which will increase dramatically in the next few decades. It will inevitably transform the society economically and culturally. What I fear will not change is the lack of understanding of the phenomenology of old people who know, as younger people do not, that with each passing year their end is nearer. Total care facilities are centers for endings. In these centers, it is next to impossible to go through a day and not be reminded that the days are numbered. Old people do not create, develop, and administer these settings. Younger people do. For young people to identify with old people and understand them, their feelings, needs, and sensitivities is no easy task. Old people are seen by younger people through different prisms from those of old people, just as in the case of parents and their children.

The future cohort of old people will be very different from the present one: More residents will be among the most educated cohort of old people and they will have lived through a post World War II social change during which the country underwent a speedy social and cultural upheaval. I cannot predict what the consequences may be except to say that it is most likely that their reaction to institutional authority will not be as conforming as that age cohort is today.

The second reason I told the story is taken up in the next chapter in which I employ a way to sensitize readers to the predictable aspects of the phenomenology of old people who enter a total care experience. Younger people have found my comparison helpful.

Chapter 3
Residents as Immigrants

We acknowledge that America is the land of immigrants. In the nineteenth and early decades of the twentieth century millions of immigrants came to its shores. Some came because they wanted to, but most came because they had to escape persecution, poverty, or both. To all of them America represented a new lease on life, free to live their lives in unaccustomed freedom. The "old country" would not be like the "new country," which welcomed and needed them in a fast developing industrial society. They came with hope, great expectations, and mixed anxiety about the unknown, a total unknown. They had been part of a community of people which was their nuclear and extended families. They had no intention of ever forgetting their relations and their community. The old country was in them in the long, arduous, dangerous voyage to America. They were plagued by memories of those they left behind and even regret that they chose to separate from them. They felt alone and lonely.

"How will I find somebody from the old country who can help me?" Aside from those emigrating from the British Isles, immigrants could not speak English. Some of them had the name of a countryman who was already in America, but who was in Chicago, Detroit, Minneapolis. How do you find out how to get to these places if you can't speak English? Stories abound and I heard some in my youth from my grandparents, aunts, and uncles. Some religious, ethnic, and philanthropic representatives at the port of entry were helpful. To the new immigrant they were proof positive that there was a God. It was a hellish experience for all of them; a destabilizing one, a source of tension among members of the family. For the emigrant, leaving the old country and surviving in the new one would only be justified if they could be with their own kind. They felt alone, lonely, and unprotected. What was balm for their tortured souls was the fantasy that their problems would soon be over. Even within the confines of a community of people of their own kind who were struggling for survival, not existence but literally survival, in the face of illness and poverty—the new immigrant was largely on his own. Housing and sanitary conditions were abominable, sometimes worse than in the old country to which some would return if it were financially possible. When everyone is coping with the same overwhelming problems, they are not primed to want to listen (if they had the time after a twelve or more hour working day, six days a week) to the woes of others.

But, ironically, despite the experience I described, different clusters of immigrants from the same country had an ever-present fear of the death of a parent

S.B. Sarason, *Centers for Ending*, Caregiving: Research, Practice, Policy,
DOI 10.1007/978-1-4419-5725-2_3, © Springer Science+Business Media, LLC 2011

or child: They wanted to make sure they would be buried in the same cemetery of their fellow immigrants according to their religious traditions.

As I was writing the above paragraph, I had a fantasy. It was my practice when I was teaching graduate students in psychology to ask this question: What are you assuming about today, this very day, without which what you are or will be doing today makes no sense at all? No student ever gave the correct answer: "I will be alive tomorrow." What if I asked that question to those long-ago immigrants, many of whom, say, were 20–25 years of age? My untestable prediction is that many of them would have given the correct answer.

We have been living in a youth culture. Many TV and radio stations target their programs for those below the age of 45. The baby boom was one big boom. Soon after World War II, towns and cities were forced into a school-building program to accommodate a number of students for which they had not planned. I know of one instance where a new school went on double session when the school opened. We are at the very beginning of a senior citizen boom for which the country is not prepared. Total care facilities are proliferating at a fast clip, but not fast enough for the millions of aged people coming down the road.

Those who enter these centers are immigrants. They left the old country for a new one. If, as I have, one asks these immigrants if they are content to be where they are, he gets many "yes–but" answers: Yes, because of problems of health, death of a spouse, living alone in their home, and a shrinking network of friends and relatives, and in more than a few instances they wanted to be nearer to their offspring. Rarely do they give only one reason. One resident said, "Am I content? Yes, because I am where I have to be. But am I happy that I am here? Definitely not. I wish it had worked out that I lived out my days where I was. It is a rare day when I do not think about it."

I am not suggesting at all, of course, that residents of these centers have experienced what immigrants of long ago did. These residents are financially independent, they were and are able to make choices, and they were free and independent to a degree unimaginable to "real" immigrants who had to cope with a language barrier.

The point of the comparison is not to suggest that one group had a harder time than the other, but that immigration of any kind comes with a price, a psychological and social price.

Visitors to these centers are much impressed with the setting, the physical appearance, facilities, and services, and residents who appear to be content in their final home. I, like other residents, do not deny that such an impression has some validity. What I do assert is that such a comparison prevents recognition of dysphoric concerns, which are ever present in the minds of the residents when they leave their "old world" to come to a new one where they know there will be daily reminders that they do not have long to live. If they know this, they are loathe to talk about it, which has the paradoxical effect of increasing their feelings of aloneness and the fear of a short-lived future, which may include debilitating illness. One resident said to me, "It's easy to die, it's hard to live."

There are two groups who have different experiences when they come to the Center. The first group is somewhat smaller than the second. The first group may or may not have a spouse, did not live in the New Haven area, more than a few come from distant states, they may or may not have a son and daughter in the area but the immigrants know no one in the Center. Uppermost in their minds is where and how they will meet residents with whom they might be compatible. And by uppermost I mean that they want soon to go to bed at night knowing that they met a person or couple who were very friendly and offered to be of help to them. That, they find, is no easy task that can be resolved quickly. How rewarding friendships are formed and sustained is near countless, but in a center for endings, they are not numerous. The entering resident soon learns that many residents have few if any serious interests in anything. Residents somewhat feverishly try to arrange to have dinner each night with people they like; they are a group and the new resident has few or any opportunities to become part of it. They need not to be alone; to be part of a group should never be underestimated. The entering resident who is a devoted scrabble, poker, and especially bridge player has a somewhat easier time meeting making friends.

There is an interesting division at the center between two groups, each containing men and women. The much larger of the two groups is comprised of residents who fill their days with activities bearing little or no resemblance to those activities in the "old country." That includes, of course, women who were traditional homemakers. That was their work, insuring that their days were full. They are not homemakers in the Center and for some, that is no cause for regret. So how do they fill their days when they enter the Center? It is my distinct impression that that question is a nagging one which preoccupies them daily and causes them to go to any sched-uled event even if it is not the one in which they have an interest; they know it is highly likely it will not be a rewarding one. They will go to the live concert on Sunday afternoon even though they do not listen to classical music. They will go to scheduled lectures because it is something to do and not because the topic is of interest to them. They will flock to the community meetings where they sit passively listening, never asking a question. There are a few women who do ask questions or offer criticism and/or suggestions and, without exception, are individuals with much education who had work experience or had engaged in local community affairs.

The men, of course, are a decided minority who have been academics or had held responsible positions in private or a non-profit organization or in clinical practice. If their number is small, so is their level of participation in lectures or community meetings. It is my impression that, relatively speaking, few of the men participate in these open forums unless the subject matter is vital to their interests, e.g., the plan to enlarge the Center.

I do not present those observations with a secure feeling. As I said at the begin-ning of this book, I did not enter the Center to study the people there. For one thing, I was far more preoccupied with my own personal and health problems. For another thing, it took a year for me to feel "at home" in my new country. Centers like mine have for all practical purposes not been studied with the seriousness and depth they deserve. And by serious, I mean developing procedures and relationships which go beyond observation of overt behavior and illuminate what people think, ruminate, and fantasize about. They are certain that death for them is in the near future. If I can say anything with a sense of security, it is that residents rarely or never put such thoughts in words for anyone to hear. Let me give two examples.

A new resident concluded after her first year at the Center that there were residents who had severe personal problems in regard to health and future for which the Center provided no helping service. She discussed this with a local clinical social worker who said she would be willing to lead such a group to help them learn to see that their problems were not unique to them as individuals, and to discuss alternative ways to cope that were more adaptive. You could put it this way: misery needs and loves company and the company can be creative and supportive in changing how you cope. In putting it that way I am not in the least being critical. The idea is a variant of the rationale of Alcoholics Anonymous. The participation of residents would be paid for by Medicare. No such group could be formed. The barrier against talking about a foreshortened future is very strong. Nothing was said about death and dying in a memo circulated to the residents. I am sure that residents vary considerably in how they think and cope with the knowledge that they may die tomorrow, next month, next year. Death is the 900 pound gorilla resident in the Center. You never, but never, would know that by what residents talk to each other about. That was true, but far less so, in the "old country" from which the residents emigrated." In the course of an average year, 17 residents die either in their apartments or in the nursing home. That is 12 percent of the total number of residents.

To younger readers of this book it may seem that I am painting a picture of residents in these total care facilities as obsessed with death. That is not my contention or the case. What I have asserted is that no day passes when the thought of their death has not been present or has been reminded of it by the death of others. The strength and consequences of the thought vary considerably and it is most consequential when they envision that they will die alone, no one who loves them will be at their side. Very aged residents may no longer have friends who are alive, they may even have outlived their offspring or those offspring may be scattered all over the map and have responsibilities which prevent them from visiting for more than a short period of time. Modern society has wreaked havoc on the nuclear and extended family. It is understandable if the thought of the possibility that they will be alone when they die is so feared by many residents. There were many immigrants in bygone days who, male or female, were not married. They were young, relatively speaking. Uppermost in their minds was to find some one or group from the old country who would guide, inform him or her about what to do and not to do in this strange new country with its strange ways. What if he or she got sick and, God forbid, could not work? What if the illness meant that he or she might die? Who would be there to help care for him? Where and who would bury him? Who would let the people back home know what happened to him in the densely populated urban ghetto with its unsanitary conditions? The mortality rates among children and adults were sky high. If you were sent to a hospital, it was a sign that you had a fatal illness.

When I was 14 I came down with polio in both arms, nose, and throat. I was taken in an ambulance to a hospital for people with infectious diseases. My mother was not allowed to accompany me. I was put in a glass enclosed room. My parents could see me through the glass windows and they could visit for brief periods of time on certain days of the week. I was in the hospital for 2 or 3 terrifying weeks. I felt unimaginably alone and when they did a spinal tap without any anesthesia, I was certain I was going to die, whatever that meant to me at the time. I trust it will be understandable to the reader when I say that I have been a first class worrier when I think that someone I love may be seriously ill and may die. Death, separation, aloneness are inextricably linked in my mind from my earliest days. In fact, that trend is number one on a list of factors that are ranked for the degree of stress they cause. Admittedly, I may be an extreme case, but I know of no data to contradict my assertion that trend gains revived strength in old age, however masked or controlled in one's earlier years.

When I was 17 my grandmother became seriously ill. She and my grandfather had emigrated from Russia early in the twentieth century to escape being killed in the pogroms. They came over with their eight children, four daughters, and four sons. My grandmother never learned to speak more than a few words in English. I never heard any English spoken by anybody when my grandparents were around, only Yiddish. I loved my grandparents.

One day when I came home from class at the local college (University of Newark, now a Rutgers' satellite campus), there was a note for me immediately to come to my grandmother's apartment. When I got there, all the daughters were there, sobbing, crying, wailing in a room adjoining my grandmother's, whose door was closed. I was told that my grandmother had a massive stroke, had become comatose, and her end was near. I learned that each daughter had taken turns sitting at her bedside but after a couple of minutes came out and began to cry and wail. They could not control themselves and, yet, in between sobs proclaimed that their mother must not be alone in her room, it would be a sin to let her die alone. What if only for a few seconds she awoke and no one was there?

Frankly, I do not know what prompted me to say that I would be at my grandmother's bedside until one of my uncle's arrived. I had been to a fair number of family funerals, would keep my gaze away from the coffin, and if it was an open casket I would be frightened and would not be able to look for a second at the dead person's face. I had my share of dreams of some impending doom.

For another facet of the theme to this chapter, let me return to the story I related in a previous chapter where I told how the residents of my center reacted to the plan to enlarge the center where I reside. When the plan was presented to all the residents and the question and answer period began, there was an initial reaction of resentment about one aspect of the plan compared to which other reactions were voiced with little emotion but rather with sheer curiosity. *That aspect was the placement of the nursing home at the far end of the Center, a fact that would make it difficult for residents to visit friends.* It would be too tiring, especially if they had to use a walker or wheelchair.

Why such obvious, quick, articulated resentment? Like all expressions of deep feelings, it contained, implicitly or explicitly, more than a message. Remember that the residents have seen friends who may have had to be moved from the independent living section to the assisted living one or, more likely, to the nursing home. Not all who go directly to the nursing home will remain there until they die. There will be those who go there after a hospitalization which then requires a stay in the nursing home before returning to their apartment. These stays vary in length. (Mine was 2 months.) Visits from friends may vary from one to several days a week. In the case of married couples who have a spouse in the nursing home, it is frequently every day. On weekdays the nursing home has a steady stream of visitors. You do not have to be a sage or a psychologist to intuit that when residents voiced resentment about the placement of a new nursing home, they were also saying, "When I will be in the nursing home, I want to see my friends and I want it to be as easy as it is now to visit me. And if when my days are literally numbered, I want to see them."

There are more than a few residents whose sons and daughters live a long distance from the Center and who need and want their friends at the Center to be with him or her when the end is near; they do not want to be alone. Yes, the residents were directing their criticisms to top officialdom of the Center because it reminded them of their own dysphoric fantasies about their final days.

It is instructive and appropriate that I acquaint the reader with what occurred after the first meeting when the plans for enlargement were made known to the residents.

Shortly after the first meeting the director held separate meetings with residents on each of the six floors of the Center. The residents were not reassured. The director spoke for most of the hour.

Two months later the residents received the following memo from the director of marketing.

Throughout the month of July, there were nine meetings of residents, staff, and project architects held to discuss various aspects of the project design. Topics included Assisted Living, Skilled Nursing, Landscaping, Dining, Community Common Spaces, Administrative Offices, and Operations. The purpose of these meetings was to ensure that the architects have an adequate understanding of each topic in order to define a valid footprint, that is, the size and location of each of the proposed building additions, in preparation for the zoning process.

The participation of and interaction among residents, staff, and architects proved extremely valuable in identifying and resolving a number of design issues. The architects have documented all the input from these meetings and will incorporate suggestions into the design. The next step is to prepare a set of schematic drawings appropriate for presentations to the various municipal bodies associated with the zoning process. The drawings will be prepared during the month of August so as to meet an early September start date for zoning. At that time, documents will be reproduced and distributed for the benefit of all to more closely examine the design.

Once the zoning process is underway, the design team meetings will resume—probably in mid to late fall. At that point, we will again schedule meetings to resume the discussions started last month, initiate dialogue on new topics, and develop even more detailed plans. This iterative design process will continue well into the spring of next year, until the architects have all the detail they need to develop construction documents.

Thank you to all the residents, staff, and Board members who participated in this exciting process. Over 30 different residents attended these meetings, representing 10 resident council committees. The perspective was broad and the dialogue constructive; consequently, we are headed toward a much better design than would have been achieved without their engagement.

As we focus on zoning for the next few months, things may seem a little quiet. But, we will publish drawings and notify you of the zoning process as soon as we have the information in hand.

We look forward to your continued assistance in developing this project.

I find the memo interesting in several respects.

1. Why is the memo from the marketing director and not the top director? The president of the resident's council is no brash activist or radical, but someone who will avoid any open confrontation with the director or board. That is why I was so surprised to hear her express her disappointment, bordering on anger, at the bulldozing style of the director. More than that, she was going to tell him and the Board that if the conflict between the residents and the director grew more heated, or became known to the public, the Center was in jeopardy. In any event, I am of the opinion that the fact that the memo came from the marketing director was not happenstance. The memo is an effort to repair communication by saying that residents have a vital role in planning, their ideas and suggestions are important, they should continue to participate "to insure that the architects have an adequate understanding of each topic." What is left unremarked by that statement is why that was not done much earlier in the long planning process. That is not a trivial question because the exclusion of residents from the planning process is the decision of the director and the advisory board. Architects are not disposed to take a stand with those who hired and paid them. The exclusion of residents bespeaks of attitudes and understandings of old people that are superficial, demeaning, disrespectful and, I must add, ignores the courtesy implied by the democratic ethos. Winston Churchill once quipped that democracy is the worst form of government except for all the others, by which he meant that it puts many constraints on leaders to have their own way by their own time perspective and without challenge from those he regards as uninformed and ungrateful.

2. The memo is very upbeat about the progress made in the recent meetings. And yet, there is not a single mention of any decision, tentative or not, that met with approval. Presumably, at some point the residents will be informed. The immediate task is to prepare a presentation to the local zoning board, one of several local and state agencies who must approve the building plans. But I did learn from a resident who was at the recent meeting about two decisions that were made. I will discuss each separately.

3. Because of the shape and size of the land on which the present Center was built, it is impossible to place the new nursing home other than where the residents criticized. The new nursing home will be significantly larger than the present one. Conceivably it could be placed elsewhere but only if it was unconnected to the other buildings. That would be intolerable. If true, that was never fully explained when the project was first presented to the residents.

4. The present dining room is a very crowded place at dinner time when waiters and waitresses serve you the meal you selected from a printed menu. There will be no such service in the two new buildings. Each will have an elaborate buffet from which you select what you want and carry it to a table. Every resident will have the option each day to eat in any of the three dining rooms. Question: how

will residents who use a walker, or are in a wheelchair, or rely on a cane manage an elaborate buffet? In the present dining room there is a small section for those who prefer a buffet: it is a small section, it has less options from which to choose than in the rest of the dining room. The buffet section is usually full. I like buffets but I have one usable arm and I do not enjoy making frequent trips to the buffet. One plate at a time is a nuisance. There was a period of weeks when I had to use a walker to go to a buffet. I do not want to give the impression that a number of people with motor handicaps is large among those capable of independent living. That is clearly not the case. So, when the enlargement process is completed, is it not likely that handicapped residents will have less opportunity to eat with friends who are not handicapped? There is no institution which has to be more "user friendly" than those built for old people.

One more example. In the middle of the top floor (sixth) is a section which would make a most attractive penthouse. That middle section contains well-appointed conference rooms and a couple of guest suites. But most of that middle section is a room used for community meetings, movies, live concerts, and special events. For people with normal vision and hearing, the acoustics leave a lot to be desired. For most old people, age has taken its toll on their visual and hearing acuity, and for many the acoustics are a source of difficulty and frustration and some, like me, attend events there infrequently. Obviously, whoever designed that room had good intentions but was ignorant about the characteristics of those who would come to it. We know from eating in restaurants that there is a difference between the appearance of the food we ordered and its taste. Architects rivet on appearance of the structure (internal and external) of the building they are designing. We are told by them that form follows function. Old people, in or out of an institution, are not homogeneous on a whole host of variables. But precisely because they are old there are certain psychological and physical characteristics they have in common and if you have not made a serious effort to learn what those characteristics are, it reflects a combination of ignorance and insensitivity.

Thus far I have talked only about those individuals who were considered capable of independent living. If you were to spend two successive days in a total care facility, you will conclude that a significant number of residents are likely in the near future to require a degree of care available only in the nursing home. They are frail, use a walker, or are in a wheelchair. Whatever they do, they do slowly and they tire easily. Many have visual and auditory problems. Some of them have already had a stroke from which they recovered after a stay in the nursing home, others have had heart attacks. Most residents in independent living are "normal" in appearance and speech but many have a history of serious illness which is likely to recur and require placement in the nursing home in the relatively near future. My answers to these considerations are contained in the next chapter where I shall discuss my experience in the nursing home. It is in almost every respect, unfortunately, like all other nursing homes I have been in as a patient or as an observer.

It was my intention to devote the next chapter to my 2 months stay in the center's nursing home. However, when I read and re-read this and the previous chapters, I realized that I had omitted things about me which would help explain how I came to understand many of the residents as well as my judgments about centers for endings. Those omissions are also important if the reader is to comprehend why my stay in the nursing home was so upsetting and infuriating. There are clear differences between spending your days in your independent living apartment or in a bed in the nursing home. But when the two are part of one institution, they inevitably have features in common.

Chapter 4
Some Aspects of Organizational Craziness

From the time I finished graduate school in 1942 I developed a strong interest in how organizations, any organization, worked and differentially affected those who worked in them in ways that facilitated or were barriers to the realization of their purposes. My first job was in a new state institution for mentally retarded individuals; it was there that I very quickly learned about what I came to call "organizational craziness," a craziness that shaped lives and purposes of *everyone* in the organization for good or for bad. The history of human service institutions is a grim one and very unpleasant reading. That history confirms the adage that the more things change the more they remain the same. Initially my interest in organizations was a matter of self-defense: I was being influenced, shaped, and pressured to think and act in ways incompatible with my beliefs and values. For the rest of my career up until a few years ago I took advantage of every opportunity, and I had many, to observe and consult to human service institutions. I learned a lot and the most important thing I learned was that organizational theorists and practitioners in the private sector were learning the same thing.

With that introduction let me start with a problem I confronted soon after I came to the center where I now live.

My apartment had been two one-bedroom apartments. Now when you enter the apartment, you see a small open kitchen on the left and the rest is a dining and living room area. You then see a hall, at least 20 ft in length off of which is a modest sized room, two bathrooms and a very small room containing a washer and dryer. At the end of the hall is a large master bedroom. It is a spacious apartment overlooking woods and a lake. The two bedrooms and two bathrooms have emergency cords you pull in case you are in need of help. I noticed that there was no emergency cord in the kitchen and living room areas. What if you fell in that area? How do you get to an emergency cord? I thought it strange but said and did nothing.

A few weeks after I became a resident I was in the far bedroom when the phone rang. It was a friend I had asked to visit. He was in the hall, had rung my bell, and was puzzled that I had not come to the door. He was calling me on his cell phone. That was when I learned that when the doorbell rang, it could not be heard in the far off bedroom. And in the event that I had the TV on, I could only hear the bell if I was in the kitchen.

S.B. Sarason, *Centers for Ending*, Caregiving: Research, Practice, Policy,
DOI 10.1007/978-1-4419-5725-2_4, © Springer Science+Business Media, LLC 2011

To make a long story short, over the next few weeks I made calls to various department heads who said they would look into the matter and call back. Nothing was done. I then went to the marketing director whom I had come to know. I told her that if nothing was done, I would notify the local and state department agencies. She spoke to the executive director and I was told that something would be done. Two weeks later a member of the maintenance department came and installed a separate bell near my bedroom. When activated by the doorbell, a bell would ring near my bedroom. It did activate the bell which made a pipsqueak noise which you could not hear if the TV was on. The maintenance man told me he was doing what he was told to do.

My resentment and stress increased day by day and then week by week. I did not want to be seen as difficult. I did not want to antagonize staff upon whom I was dependent and with whom I had to live in my final years. This reluctance to complain, to antagonize, to stand up for what you believe is fair and right is characteristic of all residents I know, especially women. They are dependent on staff in numerous ways. They do not grin but they bear it, poignantly aware of their dependency.

What was the end of the story? *Well, the end of the story took place over a year later and only after I threatened legal action.* What stimulated my threat was being told by maintenance workers that I was absolutely correct about the problem but they could not get approval to do more than they had done. *They feared antagonizing the top executive and other top administrators.* From all that I have learned from a large number of employees, the top administrators operate by fiat and deadlines. They assign tasks and the deadlines by which they must be completed. If the deadline is not met they will not listen to excuses, you are given a negative evaluation which can lead to being fired or no increase in salary.

There may be readers asking why I am telling them about the quality of relationships between employees and administrators. The answer is that in the case of human services institutions the quality of that relationship inevitably is reflected in some way or degree in the relation between top administration and the residents they serve. For example, over the year and a half that I tried to get top administration to fix my doorbell problem I must have phoned them 50 or more times. In all but three or four of those times I had to leave a voice mail message with the request that they call back. I am, like most people, the kind of person who takes the rules and obligations of courtesy seriously. If I leave a phone message on Monday morning I expect a return call later that day or on the next day. That, I learned, was an expectation very rarely satisfied. I discussed this with other residents most of whom had similar experiences and feelings but unlike me had never voiced their resentment to the director or top management. Why didn't they? My answer (= opinion) is based on two notable features of the meeting at which residents were first told about the plan to enlarge the centers I discussed in an earlier chapter.

Prior to that meeting the gossip mill had already alerted the residents that planning the enlargement had been going on for 2 years, a source of anxiety and resentment for the residents. When at the meeting the implications of the enlargement became clear, they were visibly upset: And yet only a handful of them stood up and voiced criticism and resentment. The rest were silent.

At dinner that night only anger and resentment about the plans were what was expressed.

Aged people come to these centers in order to continue to be independent, capable of planning their final years in ways that will be satisfying. Moving to such a center is literally a moving experience because they are giving up an accustomed way of living which for one or another reason was burdensome. They come with the expectation that they will be provided with services appropriate to their needs, the most important of which is to feel independent, to be respected, not someone devoid of opinions and ideas which should be heard. They did not come with the expectations that the management of the center would take any action, major or minor, that will affect them negatively unless the resident has some input to the plan of action. It is what I call the political principle: if you are going to take an action which will affect me, I do not want you to take action unless I have the opportunity to present my views.

People in positions of power are the most frequent violators of the political principle. They make decisions affecting people with less power and then inform them of the decision. In the case of the center's expansion plan the executive director violated the political principle, not in his mind as an exercise in power but in the belief that Papa knows best and the rest of the family should get behind and march. It could not occur to him that he was treating the residents as if they were too unsophisticated and inexperienced to make a contribution or, worse yet, that they had legitimate self-interests that should at least be heard and discussed.

I did not come to the center with any expectation whatsoever that I would want to do battle with institutional authority. But if I had such an expectation it would be accompanied by the knowledge that challenges to institutional authority stand little or no chance of being effective unless it was by an organized group, not by an individual or two or three individuals. The power of the residents is in their numbers, their willingness to take a stand, and the emergence of leadership. Whether the numbers are large or relatively small as my center is, there are several barriers to action.

1. There are residents—their number not miniscule—who are loners who have been unable or unwilling to become and feel part of at least one small group. In earlier pages I described my attempts to meet residents in my early months in the center. It was by no means a rewarding experience. Why should it be rewarding? Friendships have to be developed, the spark of mutuality has to be ignited and nurtured, self-revelation has barriers to overcome, and the odds that strangers will become friends are far less that they will have a basis for friendship. When in addition you take into account that aged people in these centers have been psychologically and physically severed from their past, that they know they are not "what I used to be, that I know the future is not one of sweetness and light, given my physical symptoms"—these stances bring with them the fear that the need and search for friendship may bring disappointment and rejection.

2. To take a stand against institutional authority requires self-assertion or what we conventionally mean by courage. That is true for action by an individual. It is

even more true in centers for aged people when collective action will be required. For one thing, they know they are dependent on institutional authorities for needs and purposes that led them to enter a center. They literally fear that challenging authority will result, directly or indirectly, in retaliatory actions. Whether that is a rational or irrational view is quite beside the point: *dependency is a breeder of fear and an inhibitor of action. To the aged residents that fear is given force by the fact that they have no options.* They have paid a lot of money to enter the center, the monthly service charge is the opposite of small. Even if they could afford it, and a large majority could not, they cannot contemplate moving to another center. One resident put it very succinctly. She had been a resident for 10 years and had lived through the turmoil and inconveniences of building new apartments at each end of the main building. "When we were told that the new expansion would take 4 years, I was enraged. It was like being in a marriage gone sour and divorce was impossible."

No leadership emerged to coalesce the residents in expressing their resentments. For perhaps 10 minutes I gave thought to exercising some leadership. I am no shrinking violet and I am allergic to leaders who believe they are God's gift to orga-nizations which need improved efficiency and who are, to put it the way I think it, cold egotistical characters who cannot imagine that anyone below them, especially residents, have an idea worth listening to. I have had only two one-on-one meet-ings with the director and I have seen him conduct meetings. He does not interact with residents in any personal way. He is for all practical purposes, invisible. It may sound unbelievable but in the 3 plus years I have been at the center I have never seen him in the dining room or in a conversation with a resident. In fact, he is such a textbook example of the unilateral top–down style of leadership that I made it my business to get to know several employees to check the validity of my assessment. I told them I needed their help because I had formed an impression of the direc-tor which I found disturbing and wanted to check it out with them. They agreed with me. I told each of them that if I was right it was not because I was a sage but because I had observed and consulted to human service institutions (and agencies which have, unfortunately, too many top executives like the center's director).

What about the board of trustees the members of which are from the city's metropolitan area? Where are they in the total picture? What responsibilities do they have and on what basis do they approve plans and policy? I have never met or talked with a member of the board and neither has anyone else with whom I talked. The only possible source of information for them is the executive director. The presumed virtue of a board consisting of individuals from local communities is that they will have more opportunity, time, and willingness to get to know the facil-ity, its residents, and employees. The great majority of these facilities are part of a private sector corporation with headquarters a distance away. They can only know each facility in a superficial way. In fact, in recent years the mass media plastered their pages with accounts of private sector boards of trustees each member of which did not have a clue about what management was doing.

I cannot refrain from relating an event which conveys aspects of what I have been saying. It was the annual cocktail hour party to which all residents are invited to meet and talk with the board of trustees. It was held in a very large common room used for movies, concerts, and lectures. At each end of the room are two couches and several arm chairs. The rest of the room contains about a dozen rows of chairs. For the cocktail party the chairs had been removed.

As I approached the room I was surprised to see residents leaving it. When I entered the room, all I could see was a mass of standing people. People were sitting on the two couches and the several arm chairs. I could not see the far side of the room.

Why were people leaving the room minutes after the party began? Why did I leave after 10 minutes? *The answer is that there are a lot of aged people who, like me, cannot stand for a long period of time, and for whom 15 minutes is a long time.* I later learned that no more than half of the board was there.

What is the significance of that event? Significance is in the eyes of the beholder. If you were on the board or in top management, you *probably* would view it as an event informed by a democratic ethos, as if to say, "We are all in this together. We have your best interests at heart. We know and respect you and we want you to know that. We know and appreciate that you know that our first responsibility is to meet your needs. That is what it is all about. That's what we celebrate today."

To residents like me, and they are not small in number, the cocktail party was another instance of officialdom's ignorance or insensitivity to, in this case, our physical infirmities as well as our desire to get to know those responsible for policy and planning. To get to know somebody requires more than standing around in a very crowded room talking to a stranger and soon parting as strangers. The end result is that the we–them divide is never bridged. The cocktail party was a charade in which the appearance of good cheer masked the internal reality of the absence of community.

If you cannot judge a book by its cover, it is no less true that you cannot judge an institution by its physical appearance. You would be correct in saying that that caveat is a glimpse of the obvious. You would also be correct in saying that it is true for understanding a family. Part the curtains of the overt appearance of a family and you soon learn that each of its members covertly sees and judges each other in ways that surprise you. Overtly the members appear to be happy, content, and expressive of their inner feeling for one another despite differences in age, gender, and temperaments. But what if you lived with that family for, say, a few weeks and were able to establish a relationship with each member and get to know them, him or her and their parents? Do you have any doubt that at the end of your stay your understanding of that family would be very different than it had appeared to you initially? That was dramatically demonstrated a couple of decades ago on PBS by filming the Loud Family once a week for 10 weeks. It was seen by millions of people and discussed in the mass media.

Yes, the point is a glimpse of the obvious and one should bear that in mind when viewing any event of interest to you in a human service institution. Surface appearances and covert psychological realities are not highly correlated. The cocktail party was one such an event.

Let me relate one "small" event which occurred while I was writing this chapter. It was the day after the mid-term 2006 election. I entered the dining room for dinner, I saw two men I knew eating at a four chair table and I joined them. I asked them how they felt about the sweeping victory of the Democrats. The conversation was lively. Another resident approached our table and asked if it was okay to join us and we said of course. I had met him only once before and found him spontaneous in expression of his feelings. After he sat down I said to him, "We have been talking about our reactions to the election. What were your reactions? He looked at me, smiled and said, "Thank you for asking me to join the discussion. Last week I joined a table with three people. During the entire meal none of them directed a question to me. I ate in silence. When they finished dinner, they said good night and that was that."

I could identify with him although I had never had such an extreme experience. But it did remind me of several times I had dinner where no one seemed to want to talk, responded to my questions with few words, and never asked me a question. I can report but I cannot explain these events. What I feel justified in saying is that they confirm a conclusion I had earlier come to about the features of interpersonal relationships among residents. Why that is so (if it is so) cannot be determined by these and other anecdotes of overt behavior. I did not come to the center to study residents and the center. *I am sure that young readers of this book have heard old people say that the most satisfying friendships that they have formed were when they were young and that it is rare in later life to have such friendships again.* Why that is so is not because of our genes. Geographically speaking, ours is a highly mobile society. If our parents or grandparents emigrated to America, we natives have emigrated from one part of America to another, the number of times can boggle the mind. Moving is a disruptive experience, it affects families and friendships, just as it did for our forefathers. Is it that when aged people make that "final" move to a total care facility, they look forward to new friendships but they quickly learnt that the social context does not contain the conditions that can dilute the feeling of an unwanted privacy?

There is much we do not know about the phenomenology of aged people. Research has focused largely on what for brevity I will call medical problems. There is pitifully little research on what I call the interaction between institutional appearances and the covert psychological realities of residents. Have we little or nothing to learn about that interaction? Is it unlikely that what we would learn would have much relevance for how managers, nurses, aides, and geriatric physicians are selected, educated, and trained? Is it that there is no good reason to expect that what we may learn will not be important in requiring political officialdom to descend from the heights of abstraction, empty rhetoric, and bromides?

Facilities for the aged are not unique institutions; they are just different in their history, traditions, types of personnel, and the people they serve. But their institutional culture reflects a base rate of organizational craziness that is never near zero. For more than 50 years I worked in and studied efforts to change and improve public schools in order to understand why schools have been intractable to change. (For the interested reader I have included a few of my books on this subject in the bibliography (Sarason, 1996, 1995, 1990). Suffice it to say here, facilities for the aged are remarkably similar to schools in self-defeating ways which make a mockery of their public mission statements. In the case of schools the general public and political officialdom have been forced to see and confront the inadequacies of schools, and local and federal governments and foundations have literally spent *trillions* of dollars in efforts at reform. The failures have been incontrovertible, to the point that the reform of schools has become at or near the top of issues in national presidential campaigns. You cannot say that about the issues surrounding the aged now and in the coming decades, which is one of the reasons I wrote this book. In the next chapter I take up my 2-month stay in the nursing home. It is a chapter that could have been much longer than it is if I had included all that other residents told me about themselves in their final years. I did not write this book because I wanted to "prove" anything but rather because I felt compelled to give voice to what I experienced. And I did so consistent with what a rabbinic sage said centuries ago:

> If I am not for myself who will be?
> If I am not for others, what am I?
> If not now, when?

Chapter 5
Two Months in the Nursing Home

Where I reside the nursing home is called the "Health Center," a euphemism that obscures the fact that a large majority of patients will be there until they die and they know that. It always reminds me that during World War I sauerkraut was called liberty cabbage. And it was around the time that undertakers called themselves morticians or funeral directors. In principle I have no objection to accentuating the positive and eliminating the negative. I make an exception, however, when the negative is the overarching concern of people! The word and concept of nursing homes evokes imagery and obligations. Substituting wellness for nursing conveys implicitly the message, "Let's not say that your days are numbered, let us talk about other things," a message every patient knows is a form of Madison Avenue advertising. Don't call it impotency, call it erectile dysfunction. Calling a nursing home a wellness center is but one of the ways illustrative of how differently patients and staff define needs. The gulf between the phenomenologies of patients and staff has different sources and the bridging of that gulf is primarily the obligation and responsibility of the staff.

My experience with nursing homes started in 1970 when I brought my father up from New Jersey where he was being taken care of by my mother and sister. He was unsteady on his feet, he fell frequently, and he had advanced macular degeneration which seriously impaired his vision. He was a very uncomplaining man, grateful for any help he received from others.

His nursing home was a relatively new and small one approved by local and state agencies. That meant there was a fully trained nurse in charge and two licensed practical nurses. There were two local doctors on call when required for residents who had no family doctor. In those early days all costs of care were paid by Medicare and, therefore, the residents were not sociologically and economically a homogeneous group. For all practical purposes those who took care of patients were aides, which means they were not well educated and most came from minority backgrounds. Needless to say they were poorly paid. That is an old story in the history of human service institutions: Those who spend the most time with patients have the least training and the lowest pay by far. (I'll have more to say about this later.)

After 1 month of visits to my father, several things became clear. First, it made no difference what time of day I saw him, he was either in his bed or sitting in his wheelchair. Second, only once had he been wheeled to a cheerful, large, recreation

S.B. Sarason, *Centers for Ending*, Caregiving: Research, Practice, Policy,
DOI 10.1007/978-1-4419-5725-2_5, © Springer Science+Business Media, LLC 2011

room where residents could watch TV (He could not see what was on the screen.) Third, when an aide brought him his lunch or dinner tray, she would place it in front of him and leave. Apparently no one told her that to my father what was on the tray was not visually distinct entities. Fourth, I received a call from the chief nurse that my father had fallen, she was not sure if he broke any bones, and suggested that I call our doctor to come see him and decide whether he should be taken to the hospital. The doctor came, examined him, and said "your father is not in pain, he may have a hairline fracture and if he did there is nothing we can or should do about it in a 92-year-old frail man. Going to the hospital in an ambulance, taking x-rays and blood tests and watching him for 3 or 4 days can be counter-productive."

I hired a practical nurse to be with my father 6 days a week, 8 hours a day. My father lived for 2 more years in his quiet, uncomplaining, grateful way. I instructed the practical nurse to read him at least the first page of his beloved New York Times whether he seemed interested or not. In some Far East countries families are expected to take care of a sick hospitalized member during the day—bringing food, feeding, and bathing, etc. To the ears of some readers such a custom will sound dangerously the opposite of modern medicine. Will it be more therapeutic than a "family" of strangers? Progress is in the eyes and experiences of the beholder.

In 1993 my wife and I were in an auto accident. She was killed instantly; I spent 3 months in three hospitals with a broken hip and my polio, dead weight arm fractured. From the moment of the accident and the helicopter ride to the hospital preparing for the operation, I was unconscious. In the course of the operation, I became semi-conscious and heard voices and felt my body being manipulated and turned. My mind began to clear around midnight to find myself alone in a darkened room. By daybreak my mind was clear and I had no way of signaling for a nurse and I was too weak to yell. I do not know how long it took before a nurse showed up. To me it was an eternity. When she came she did not introduce herself or ask me a question. I told her I was in great pain and she then moved me to a more comfortable position. That is when I learned that there was a cast from my left hip to the ankles. She told me the surgeon would be by to see me in the morning. Another eternity. When he came I was surprised to see a young man, friendly, insecure, who told me the operation was a long one, and by no means easy. He said he would return later in the day. When I asked him if my wife was in the hospital, he said he did not know. It was not until my daughter arrived after a long airplane trip that I learned of my wife's death.

No member of the hospital staff ever referred to my wife's death or expressed sympathy for how I must be feeling her loss. That's not their "business," which is hips, fractures, etc. I was in the hospital for 2 weeks, enough to know (a) that signaling for a nurse or aide meant waiting 10–15 minutes and, frequently, sending another signal, (b) propped up in bed and with one usable arm causes you to lose your appetite. It had the virtue that I was losing weight which in its wake brought constipation; (c) I lost total confidence in my surgeon who was inept, insecure, and stupid: He walked in one day with an airplane splint which would encase the upper part of my body and on which my right arm would rest.

I told my daughter to arrange for me to be flown in a special small plane to New Haven where I had friends and I know many of the medical school faculty. She persuaded me to be flown to a major Harvard-related hospital in Boston. She had a 3-month-old infant and could see me more often in Boston than in New Haven. She lives an hour from Boston. The director of the hospital told Julie that he would choose an orthopedist, in this case it would be Dr. Murphy, who was young but already recognized as gifted. I was taken by ambulance from the airport to the hospital and was, to my amazement, greeted by Dr. Murphy. When we got to my room, he told me that the operation in New Jersey was incomplete and would require another operation. He also told me that a body splint for the fractured lame arm was unnecessary, I could keep my arm in a sling for a couple of weeks by which time the fracture would probably be on the way to being healed. It worked out as he said it would. But one thing was the same as the previous hospital. Woody Allen once quipped, "Not only is God dead, but try to get a plumber on Sunday." Substitute the word nurse for plumber.

Ambulatory patients are less needy of help from a nurse or aide. But there are many patients who are needy and know they are seriously ill or whose physical movements are restricted by this or that tube attached to or in their body, and are psychologically needy for hands-on help or reassurance as quickly as possible, and by quickly I do not mean instantly, but 15 or more minutes is excessive.

Am I making a mountain our of a mole hill? Am I exaggerating the frequency of long-waiting times? Am I "picking" on nurses and aides? I shall return to these issues later in the chapter.

I was then transferred to a rehabilitation hospital near Boston. I was there for 6 weeks. For the first 3 weeks I was flat on my back unable to change my body position because of the cast on my hip. During those initial weeks I rarely saw a nurse during her shift. Aides would bring me my meals, every few days a lab technician would take my temperature and blood pressure. I saw a doctor once. To the staff I was not sick as a cancer patient is sick. I was *something* that was healing, so what was there to talk about? The fact is that in all the time I was there I was not a person, I was a broken hip.

After a time the cast was removed. For the next week my social life changed in one respect: A physical therapist came for 15 minutes to watch me exercise my left leg. After that week I was wheeled once a day to a large room in which at least each of 30 patients were being put through a set of exercises under the supervision of a physical therapist. Although it was a heavily populated room, it was a relatively silent one. It was hard for me to avoid the conclusion that most of the therapists were bored. I had three different therapists none of whom ever initiated a conversation with me. I have no doubt that most of these therapists are capable of sentiments, sympathy, empathy, warmth, and encouragement but precious little of these characteristics were ever in evidence.

In April 2006 I fell and fractured my right hip and there was a hairline fracture in my pelvis.

I must tell the reader that what I shall describe and discuss in the rest of this chapter I have presented to the center's nurses, medical director, at least a half dozen aides, and the vice president who has ultimate responsibility for the nursing home. I did not do this because conditions in this nursing home were intolerable. They were not intolerable; no one there is a villain who knowingly mistreats patients. If such mistreatment is spotted, the aide is immediately fired. No, the major problem is not a lack of good intentions, far from it. There is no one major problem but a constellation of direct or indirect problems. Rather than resort to abstractions and generalizations, I shall present actual experiences which will allow me later to discuss a truly overarching major problem of which people are unaware, and by people I include federal and state governments. That should not be surprising. The large and influential American Association of Retired People, many of whose members are in a nursing home, has been totally silent.

In the morning of the first day in the nursing home an aide came in to change the sheets on my bed. She was a strong, non-talkative, fast-working individual. To change the sheets required that I roll to the edge of the bed. I had great difficulty rolling over and she "helped" by placing her hand on my right hip. I screamed. The pain in my hip was intolerable. She said nothing, and when I was rolled to the other side of the bed, I screamed again. She remained silent. Three days later she was again on the team to take care of me and nine other patients that day. Act 2 was a replay of Act 1.

It later occurred to me that in what is euphemistically called training of aides she was told little or nothing about how to avoid exacerbating pain in fractured areas. In Connecticut, to be called a Connecticut Nursing Assistant a person has to take and pass a 4-month course. By the time I left the nursing home I had interviewed many aides about the course. I learned two things. First, for all practical purposes the course had no content relevant to establishing a relationship with sick patients. Evidence of this may be revealed in the fact that after I had been in the nursing home for 6 weeks, no aide unknown to me (with one exception) ever entered my room, introduced him/herself by name, and asked why I had signaled to the nurses' station. The single exception was a young lady. Two days before she had come to my room I had a long conversation with the center's vice president in which I criticized what passed for training of aides. I illustrated my feelings with what happened 2 days before when I signaled for someone to come to my room. She was a young woman in her early twenties. In a tone of irritation and exasperation she asked in a loud voice, "What do you need?" I counted to three and then I let her have it. "What permits you to come into my room and ask me what I need in a tone of voice that suggests I have no right to have needs, as if I signaled for someone for the hell of it. Are you unaware that all people in this nursing home are old or frail or in pain who have more needs than you can count? Did I pay a hefty sum of money when I came to the center, and a hefty sum each month because I needed a vacation? Incidentally, are you prohibited from telling me your name?"

Two days later another aide walked in and said, "I am _____ _____; How can I help you?" I called the vice president and told him of this unusual new aide. He laughed. "I had a meeting yesterday with new aides and I told them about how they should introduce themselves to residents." What I refrained from telling him was that what I had related to him was not prompted by a concern about observing the tradition of conventional courtesy but rather the gulf between how aides and patients define "needs." The next point illustrates what I mean.

For the first month of my stay I was not allowed to stand on my legs until there was reason to believe that by standing there would be no undue pressure on my fractured hip. That meant that when I wanted or needed to sit on a chair, two aides had to strap me in a hoist which would elevate me above the bed and deposit me in the chair. No aide was allowed to do this alone, there had to be another aide. So, if I were sitting in a chair and had to urinate or defecate, my aide had to find another aide to help out, and that was a problem: Aides were reluctant to cooperate because it would cut into the time they said they needed to care for their patients. They would say they were too busy. That was one reason. Another reason, I was told, was that some aides did not want to cooperate, they did not know my aide, or they were unfriendly, or they would be criticized by their senior aide whose responsibility was to make sure that the patients assigned to the team were cared for by the end of the shift. I cannot justify the validity of any of these and other reasons. What I can say is that some of my aides said they were too busy to get the hoist at that time and I would have to wait. When I told them that if during waiting I could no longer inhibit urinating I would have to pee in my bed or chair. I never had to pee in bed.

What happened when Wayne Howard was my aide? He would go to bring the hoist, if a second aide was not available, he would shut the door and do the hoist operation quicker than if two aides did it. Wayne's obligation was to what the patient needed.

I was taken in a wheelchair on a van to my retinal specialist who every 6 weeks injects a medication in the eye in which I still have some vision. After the procedure he gave me a liquid antibiotic to prevent infection: A drop of the antibiotic is put in my eye four times a day for 4 days. With one semi-usable hand I cannot put the drops in by myself. When we got back to the nursing home, I gave the tiny container and the doctor's written instructions to the nurse. She said, "I cannot do it. We can only give medication which our doctor prescribes. She is not here today. I will fax her in her office at Yale." I blew my top. I said that if I got an infection, I could be in real trouble and if that happened, I would sue the Center. She then said that she would put in one drop but no more until she got the approval of their doctor. I asked her what if the doctor was not in her office. What do I do? I was more than upset. She then said, "Calm down. Don't get so excited. Everything will be all right." To which I said, "You can say that because it is not your eye that can get infected, it's my eye that is at issue." I didn't receive another drop until late in the evening when the doctor phoned in her approval, by which time I was totally consumed with anxiety and thoughts of total blindness. My medical needs were trumped by her need to follow rules.

I can give many more examples but their significance depends on how you answer one question: what do you mean by care of aged, sick, handicapped, dependent people? We certainly do not mean warehousing them. We do not mean merely keeping them alive. Put it this way: What do we as human beings *owe* them as human beings? It is not necessary that you feel you are repaying a debt you owe to the person you are caring for. You may never have known the person you are caring for. But when you agree to care for another person, you assume you know what that person needs and that you know how to satisfy those needs. That begs the question: How is the quality and effectiveness of the judgments of the caretaker and the cared for to be determined? Who makes that determination?

Those are legitimate questions but, initially at least, they are not productive because they convey imagery of single caretakers and single patients in the here and now in real settings. Concretely, you cannot comprehend and judge any event I have described or observed in the nursing home without knowledge of how all of its personnel are selected and trained. Without such knowledge you are very likely to make the mistake of blaming the victims and, therefore, missing the trees for the forest. What you experience and observe takes place in a complicated system which has a complicated history.

So, for illustrative purposes, let us begin with the selection process by which individuals seek to receive training as a physician. (I could just as well have chosen psychologists, teachers, social workers, even members of the clergy.) By what criteria does a medical school select would-be physicians? They evaluate transcripts of grades in college, their medical aptitude test scores, letters of recommendation, and personal statements by applicants about why they seek to enter the profession. And whenever possible they will interview candidates. The selection process is long and expensive. If you, as I have, knew well and talked with professors of medicine who served on the selection committee, you would conclude that selection was a top priority for the health and reputation of a medical school. (For years the medical dean's office and two medical departments were in the same building as the department of psychology; the rest of the medical school was across the street in Yale-New Haven Hospital.) Members of the selection committee are not in doubt about one thing: they say they want students to be other than drones with very narrow interests, little interest in the humanities, and who appear to lack caring and compassion in applying medical science to the needs of patients. The adjectives caring and compassion are used frequently.

Beginning in the 1960s reports began to appear in professional journals and mass media highly critical of the increasing number of doctors who lacked caring and compassion for their patients, because they were too busy making money to take the time to be caring and compassionate. Warnings were expressed that patients could no longer be counted on to regard doctors as Olympian Gods who were without imperfections. These warnings did not go unheeded in medical schools. Well-meant but token efforts were made to sensitize medical students to the importance of treating the "whole person," not only his or her bodily symptoms.

The dean of the Yale medical school, a psychiatrist and a good friend, decided that he needed to know the answer to this question: Of all the sources of data which were rated by each member of the selection committee, which sources predicted best the applicants who were accepted? He hired a very sophisticated research psychologist to do the study. The results surprised everyone on the selection committee: If they had used only college grades and the medical aptitude test score, they would have selected 90 percent of all applicants they had selected. None of the other sources of rating (including the interview) contributed significantly to the outcome. The medical school faculty were divided in their reactions: those who were glad that they were selecting "the best and the brightest" and those (a smaller group) who saw the outcome as an indication of how, rhetoric aside, physicians felt most comfortable with ratings of intellectual–cognitive variables. The report resulted in no changes.

So what about caring and compassion? Are the behaviors, styles, and imagery we associate with these labels too mushy, sentimental, and ambiguous to be measured directly or indirectly? The field of psychometric measurement is a highly developed one. Psychology had and still has its fair share of theoreticians and researchers who have thought outside the mold and creatively have come up with ways of measuring complicated psychological processes that cannot be developed quickly and without encountering difficulties or making mistakes that have to be overcome before the procedure has a respectable degree of established validity. Do we say to medical researchers, "Your are a pack of quacks. Since World War II you have been given many billions of dollars to find a way to prevent or cure the most frequent types of cancers." We do not say that but we willingly continue to support new research. In the area of human services issues of selection and training either are ignored, glossed over, or considered of secondary importance.

There is one group in the medical community who has learned the hard way about caring and compassion: physicians who became seriously ill and had to be hospitalized or spend time in a nursing home.[1] Several have vented their rage by writing a book about their experience. Many more do not want to reinforce what they consider common knowledge in hospitals and nursing homes: When doctors become patients they are described as demanding, critical, and are grand pains in the neck. I am not a physician, I am by training a clinical psychologist who has written about caring and compassion in human service settings (Sarason, 1986). So it is not surprising that soon after I left the nursing home I began to write this book. But it was not rage that I wanted to express, although rage was in the picture, but the knowledge that in a few decades America will have the largest number of old, dependent people near the end of their life span who, like newborns, need an unusual degree of support, understanding, caring, and compassion. Yes, we need at the end of our lives what we readily believe young children deserve and too many do not get.

Two anecdotes are relevant here because they are of a kind that is numerous but never reach the light of published day. The first is about my mother-in-law who at the time was well in her eighties, her husband had recently died, and she was living alone in Brooklyn. I was able to rearrange my schedule and go down to stay with her for a week. She was ailing, running a fever. (My wife could not go because she was suffering from fibromyalgia, a muscular condition that waxed and waned and prevented her from engaging in any physically stressful activity.) When I arrived, my brother-in-law, a cardiologist, was there and informed me his mother had pneumonia which, given her history of lung problems, raised the question: Should she be hospitalized or should she be treated at home? He had asked a friend and

---

[1] See the following articles: Biro, David (2000), Kurland, Geoffrey (2002), Pensack, Robert J. (1994), Rosenbaum, Edward E. (1988), and Weisman, Jamie (2002).

colleague to examine her and render an opinion. The colleague arrived, examined my mother-in-law and the three of us went into another room to discuss options for treatment. Finally, the colleague said, "I know your mother. She is a highly anxious woman, she will be scared out of her wits. She will never get the kind of personal care she needs in the hospital. What they can do for her medically we can do and monitor right here. You know what the hospital is like for someone like her who is afraid to be alone." My brother-in-law agreed. I was surprised how readily they reached an agreement and said out loud that the hospital was no place for a woman like her. She remained at home and a week later was without symptoms.

I was relieved at their decision. It brought back memories from two or more decades earlier when she was taken to the hospital with a bleeding lung. When I got there, she was in bed, a look of terror on her face. Huddled in one corner of the room was a group of several doctors discussing her condition in low tones. They never once talked to her. They started to leave. She looked at me as if to say, "Am I going to be alone?" I stayed with her through the night until the next morning when her son came to do his rounds and teaching. To the doctors, Pauline Kroop was not an immigrant woman who took care of her infirm husband, lost a child in pregnancy, gave birth to a Down's syndrome daughter who died in a few years, and reared her son and my wife, sent and supported them through medical and graduate school; and all of this while she was at home designing clothes for rich women. To the doctors she was a bleeding lung and doctors see no point in talking to a bleeding lung. She had no peer when it came to Jewish cooking and baking. Quite a bleeding lung!

The second anecdote is about Dr. Fritz Redlich, the Yale School of Medicine dean I discussed in earlier pages of this chapter. Several years before I became a resident at this center, he and his wife had been residents. He died a few years later and his wife more recently. They had moved here from California, a fact I did not know until after I had become a resident. While he was here, there was one occasion when he spent a week in the nursing home. Not long after their move to the center his wife had to be cared for in the nursing home.

When I was going to be discharged from the nursing home, I was a psychological and physical mess who could not get from A to B without a walker. I did not want to be alone. I feared it. To say I was unsure of myself is one of the understatements of the century. With a friend's help I located and hired a young woman to be with me from 9 AM to 12 PM each weekday morning. She prepared breakfast and lunch, washed and ironed clothes, etc. She is from the Philippines, is married to an American trained in quality control of various products, and they have two children. And we talked. I learned a lot about her homeland and family. In the course of a conversation I was surprised to learn that she and her older sister had done private duty work in the center where I had been a resident. And for whom was she a private duty aide? For Fritz Redlich and his wife who was already in the nursing home. Fritz remained in his independent living apartment, moving slowly with a walker. What I learned was that the professional staff and the aides regarded him as an overly critical, interfering, arrogant old man who was not satisfied with how his wife was being cared for and they were relieved when he hired private duty aides. His criticism had little to do with medical care but the lack of sensitivity to her as a person. (Herta Ludwig was for years a featured singer with the Metropolitan Opera Company.) I do not know, of course, what the substance of his criticisms was. But what I do know is that Fritz was long a critic of the tendency to define care in narrow medical terms. Only when you become old enough to be on an "adult" ward, are you expected to keep your inner world to your private self. Nobody is interested, unless of course you are a friend or family member.

In this and a previous chapter I primarily discussed nursing homes in general and those in total care facilities in particular. I did this for two reasons: the first is that of patients in nursing homes a majority of them are not ambulatory. When they were capable of independent living, they were regarded as "residents," when they need the nursing home they were regarded and treated as "patients," and that alters their relationship to their caretakers in one very crucial respect: they have little or no power. For all practical purposes they are in a hospital where they are expected to conform to their more knowledgeable caretakers. But as I have indicated, their caretakers, the aides, are far from knowledgeable or trained. As anyone who has the responsibility to screen aides will attest, the pool from which to select is by no means large and their judgments of those they select are by no means confirmed by their subsequent performance. The screeners are quite aware that the pool of applicants from which they select consists largely of minority women who would have difficulty in gaining employment in other areas of work.

Bear in mind that many of the patients have neither the energy and motivation to criticize or complain to them; aides are very powerful people who should not be criticized because it will get you nowhere or make things worse. A telling exception that proves the rule are married couples who entered the facility in the independent living section until one of them had to be moved to the nursing home, the spouse would spend hours with the patient in the nursing home and soon was forced to become an advocate for the patient. As an advocate, he or she would voice complaints to the aide and if that was not productive, would go to the nurse at the nurse's station. If from the standpoint of the staff the self-appointed advocate was taking his or her role too seriously, the advocate was not well liked, to say the least. When I was in the nursing home, there were five such advocates. I envied those patients.

The second reason has to do with the near future: At least five relatively new total care facilities in this part of Connecticut are already adding new wings to house people who will be capable of independent living. I am not aware that any of these facilities are enlarging their nursing home. Where will they get aides, especially those who are outside the city limits and there is no public transportation and the price of gasoline will keep climbing? *I must remind the reader that we are only at the beginning of a dramatic increase in the number of people entering the category "old people."*

The future picture is grimmer than I have indicated. *I have been describing total care facilities which the great bulk of old people will be totally unable to afford.* What will happen to them when by reason of age, forced retirement, or illness they will be unable to remain in their apartment or house? Where will these individuals or married couples go? Where are they going now? How many have offspring with whom they can live? The one thing we know is that state, local, and non-profit social agencies spend a lot of time trying to find placements for many of these people, especially for nursing homes that have an available bed which may or may not be near where these people had lived. There are nursing homes that remain open because indirectly the state is supporting them. I know of a state institution for mentally retarded individuals where conditions were such as to lead a court to require the state to place a significant number of its residents in the community. And it did so by putting them in nursing homes. This is not to say that these nursing homes were as intolerable as the state institutions, they probably were not. But it is to say that they were placed where they were because there was an available bed and not because it was an appropriate placement.

I am reminded here of the work of Kenneth Koch (1997, 1970), an internationally known poet who for years was on the faculty of Columbia University. He had the "kooky" belief that people, young and old, could learn to like and write poetry. His initial effort to demonstrate the validity of his belief was with poor elementary school children in a ghetto school in Spanish Harlem in Manhattan. Almost all of them were performing well below grade on achievement tests. He published his results in a book *Wishes, Lies, and Dreams: Teaching Children to Write Poetry* in 1970. No one I knew who has read that book has been left untouched by it.

But then he decided to work with residents of a nursing home on the lower East Side of Manhattan. His description of the ambience of the facility and of the residents can be put this way: the patients, mostly old and many in wheelchairs; sat silently or dozed and there was no social interaction among them. They were people waiting to die. Koch wrote a book about how these residents learned to like and write poetry that will not fail to exhilarate and inspire readers of the book. The title of the book is *I Never Told Anybody: Teaching Poetry Writing in a Nursing Home* (1977). I cannot do justice here to the quality of the poetry written by people who had resigned themselves to a life of passivity, silence, and hopelessness.

Where do old, sick, dependent people go when they are no longer capable of independent living? I have put that question to human services professionals and policy makers. The short version of their answers is that there is a crazy quilt of uncoordinated programs that defy description. The long version would require a thick book. There is one thing about which all of the interviewees were in total agreement. The gulf between the haves and have-nots in the quality of their programs for aged people has grown wider by the year. And when I asked them—they were all working in local and state agencies—about planning for the future they guffawed because there is no planning for the millions of old people the future will certainly bring. As one of them put it, "No one listens to us. No one wants to hear our tale of woe. And more than a few of them regard us as Chicken Littles programmed to proclaim that the situation is getting worse."

There is no simple explanation for such ignorance and insensitivity. It is not a matter of thoughtless indifference or cruelty. If the explanation is awesomely complicated, one thing is for sure. It is not the first time in our national history when the type of problem I have been discussing has occurred. Therefore, in the next chapter I will briefly discuss examples from the past when the American ethos was tested and transformed in relation to what had not been considered a major "problem" or an error of omission. And central to these examples is the question: As a society, what do we owe this or that group who for reasons not of their making, have limited opportunity to engage in the process of pursuing life, liberty, and the pursuit of happiness! These words, that ethos, are no less applicable to aged people who know their days are numbered as it is to newborns who are totally incapable of helping themselves.

Postscript

The initial Medicare and Medicaid legislation was enacted in 1965 and implementation began in 1966. When President Johnson signed it into law, at his side was President Truman who a couple of decades earlier had proposed legislation for a national health insurance program. The 1965 legislation stimulated a dramatic and speedy increase in construction of nursing homes by individual entrepreneurs, for profit corporations, and a similar number of non-profit groups. Within a decade or so the mass media were reporting what came to be known as the "nursing home scandals." The most succinct way of putting it is that profitability trumped patients. In 2000, Consumers Union began issuing reports about quality of care in nursing homes. Its September 2006 report begins with this heading: "Nursing homes: Business as usual. Two decades after the passage of a federal law to clean up the nation's nursing homes, bad care persists and good homes are still hard to find." (I urge the reader to use the Internet and look up Consumer Reports on nursing homes of which there are many over the past decade and all of which are in line with what I have said in this book.)

In the following excerpts the reader should keep in mind that in 2006 the for profit nursing homes are by far the dominant type.

For this report, we analyzed the three most recent state inspection reports for some 16,000 nursing homes across the USA. We also examined staffing levels and so-called quality indicators, such as how many residents develop pressure sores when they have no risk factors for them.

The Consumer Reports *Nursing Home Quality Monitor*, formerly the Nursing Home Watch List, lists facilities in each state that rank in the best or worst 10 percent on at least two of our three dimensions of quality. By examining the kinds of homes that tend to cluster at either end of the continuum, we can make some judgments about how likely a facility is to provide proper care.

This year's list, financed by a grant from The Commonwealth Fund, a philanthropic organization, is the fifth we've published since 2000. We've seen little evidence that the quality of care has improved since then. Indeed, 186 of the homes cited for poor care on this list have also appeared on earlier lists of poor-quality homes.

In 2002, a study conducted for the federal Centers for Medicare & Medicaid Services (CMS) noted that without a daily average of 2.8 hours of care from nurse aides and 1.3 hours from licensed nurses, residents were more likely to experience poor outcomes—pressure sores and urinary incontinence, for example. "Most nursing homes are staffed significantly below that," says John Schnelle, director of the Borun Center, a joint venture of UCLA and the Jewish Home for Aging that does research on long-term care.

The CMS, however, has not recommended or adopted minimum staffing standards, a point of contention for nursing home advocates, who are pushing for them. Marvin Feuerberg, a technical director at the CMS, says officials even watered down the 2002 study's executive summary when it was given to Congress.

Instead, current rules say that staffing must be sufficient to meet the needs of nursing home residents, a standard so vague that it makes penalizing nursing homes that skimp on care almost impossible. Rules do require homes to have 8 hours of registered nursing and 24 hours of licensed nursing coverage per day. But the standard applies to all homes, no matter how many residents they have. So a nursing home with 200 residents can use the same-size staff as one with 20.

Although the number of deficiency citations written by state inspectors has increased 7.6 percent since 2003, according to the CMS, inspectors appear to be watering them down. Each one carries a letter code, from A through L, indicating the scope and severity of the violation. Citations labeled G through L denote actual harm of the potential for death. Codes I through L indicate that the harm was widespread, affecting many people.

State inspectors are now writing fewer deficiencies with codes that denote actual harm, such as avoidable pressure sores and medication errors. "We are going back to a less stringent and simpler enforcement," says a federal analyst familiar with nursing home inspection data at the CMS. "Everything is becoming a D level. Nursing facilities are going to challenge anything above a D level if it carries a mandatory penalty, can be used in a tort case, or will be publicly disclosed.

In 2000, 40 percent of all deficiencies carried a D designation. By 2005, the number had risen to 54 percent. The reason, says that analyst, is pressure from nursing homes on understaffed state agencies that find it hard to muster the resources to defend their citations in court.

The most common remedy for violations is a "plan of correction." The nursing home acknowledges there is a problem and promises to fix it within a specified period. Often the problem is corrected but soon resurfaces, a phenomenon regulators call yo-yo- compliance.

On May 30, 2006, the American Journal of Public Health published online ahead of print a one-page summary of a research article "Effect of educational level and minority status on nursing home choice after hospital discharge." The authors are Drs. Angelelli, Grabowski, and Mor.

Objectives. The movement to publicly report data on provider quality to inform consumer choices is predicated on assumptions of equal access and knowledge. We examine the validity of this assumption by testing whether minority/less educated Medicare patients are at greater risk of being discharged from a hospital to the lowest-quality nursing homes in a geographical area.

Results. The probability of African Americans' being admitted to nursing homes in the lowest-quality quartile in the area was greater; 95% confidence interval in comparison with Whites. Individuals without a high-school degree were also more likely to be admitted to a low-quality nursing home.

The nursing home I was in for 2 months is a unit in the total care facility in which I live. By the conventional criteria by which nursing homes are judged, my nursing home would get a high positive rating. But those criteria are narrowly medical, bodily, and pharmaceutical, as if the body is one thing and the mind is another. In terms of function the mind is like the frontal, parietal, and occipital lobes and other parts of the brain. You cannot see the mind although no one doubts that he or she thinks, fantasizes, experiences pain, anger, love, sadness, a variety of needs, dependency, and numbered years. As I suggested earlier, ignorance of that reality may leave many of us victims of the anaclitic depression suffered by the orphans examined long ago by Spitz (1945) and Spitz and Wolff (1946)!

You do not have to tell parents that they have two obligations. The first is that they supply nutrients to the dependent newborn to keep him or her healthy. The second is that they stimulate and nurture the developing mind and its developing potentialities. In fact, parents, year by year, become increasingly riveted in their offspring's mind as that is expressed in language and overt behavior.

The aged person is also dependent on others for interpersonal and intellectual stimulation and nourishment conveyed by words and actions reflecting interest, understanding, and respect. The aged person does not need or want verbal bromides or empty rhetoric that betray how superficial the relationship is between caretaker and cared for. Better to retreat into an unwanted privacy and play the game that a relationship exists, that they are not strangers to each other. The criteria for judging the quality of care in nursing homes says next to nothing about these issues or even about the phenomenology of aged people. There are far more poor quality than "good" ones. I spent 2 months in a nursing home that would be judged good, even very good. But the purpose of this chapter and book is to explain why "good" is woefully incomplete and misleading. I do not blame any individual, group, or profession. Generally speaking, the caretakers reflect the grip of stereotypes in our culture.

However, there is one group I cannot refrain from criticizing. I refer to the political establishment in Washington who by their inaction is setting the stage for making a bad situation worse. Worse than their inaction is their total silence about the issues I raise in this book. They talk and talk about the millions of baby boomers on the doorstep of old age, the precarious status of the social security fund, and a galactic national debt we are passing on to our children. Why was officialdom in Washington surprised and shocked to learn that the facilities for and care of wounded Iraq war veterans were inadequate and shameful and that the Veterans Administration had long been overwhelmed by the number of aging veterans from World War II, the Korean War, and the Vietnam War? I will have more to say about this in all subsequent chapters of this book. The purpose of this current chapter was to use my personal experience as a partial answer to the most important question of all: What do we owe dependent aged people?

C-Span I carries the proceedings of the U.S. House of Representatives. C-Span II carries those of the U.S. Senate. I have long been a devoted viewer of C-Span. I cannot recall a single instance where a member of Congress recited the demographic data indicating that in 2–3 decades the cohort of aged people will triple, that Congress has not confronted the implications of those data, and there is no reason to believe that it will confront it in the next few years. Well, on June 11, 2007, Congressman Mark Kirk issued the wake-up call. His sole concern was the huge disparity between the number of existing nursing homes and the number that will be needed in a couple of decades. By the time he finished my initial positive response was replaced by disappointment. Several times he said that if the present quality of nursing homes is to be maintained, we had to take action now. Apparently the congressman was ignorant of the report of federal and state health agencies that the quality of care in nursing homes is nothing to be proud of, that however you define acceptable the number of such nursing homes is small in number. Nor is the congressman aware, as he should be, that the personnel who spend the most time with aged patients are for all practical purposes untrained, undereducated, and underpaid, which explains why the turnover rate of these personnel is very high. I am sure that the congressman is well intentioned. Who can deny that action must be taken? It is a distinction that makes a difference, a huge one, when action is based on intimate knowledge of what has failed in the past and why. In this and the remaining chapters I try to put flesh on the bones of what I have just said. Even so, I readily admit that I have not cornered the markets of knowledge, wisdom, and the truth. The problems are more than complicated, vested interests are many and powerful, cultural stereotypes abound and go unchallenged, and resistance to change—individual, professional, institutional, and political—is a certainty. In the realm of human affairs we will always fall short of the mark we strive to reach. Falling short of the mark is not a secular sin. Not having a mark is a secular sin.

Am I an angry, over-wrought, contentious, hypercritical individual who is inappropriately overgeneralizing from personal experience? Am I an anxiously obsessive person when I predict that our society is totally unprepared for the demographic changes which will transform our society? Is it really the case that we have learned next to nothing from the nursing home scandals of the past, from the history of human service institutions for the mentally ill or mentally retarded and juvenile delinquents who have been warehoused? I will deal with these questions later in the book. Let me end this chapter by urging the reader to go online or to a public library and read an article that occupied one-quarter of the front page of the New York Times for September 23, 2007, and which then occupies the equivalent of a full inside page. The headline on the first page says, "At many 'Nursing Homes', More Profit and Less Nursing." The subtitle says "Insulated from Lawsuits, private investors cut costs and staff." The investigative reporter names, places, including the greed of two very large financial firms who have been buying up many thousands of nursing homes, reaping profits, and making the nursing home scandals of the 1970s seem like child's play. State and local agencies can do nothing about it. What I find monumentally ironic is that it may be the case that these Wall Street firms are the only ones who are taking seriously the fact that the post World War II baby boomers are already knocking on the door admitting them to the status of senior citizens. When in the 1930s Willie Sutton was asked why he had robbed so many banks he replied, "Because that is where the money is." I have no basis whatsoever to call the financial firms robbers. They are business people ever alert to where the money is. Finally, let us not gloss over the fact that when in 1965 the Medicare legislation gave rise to a dramatic increase in nursing homes, neither the federal government nor health professionals asked or answered the question: What do we owe sick, frail, people. We have still neither asked nor answered that question and that is why aged people will continue to be treated in nursing homes and hospitals as if they have no minds, hopes, feelings, fantasies, yearning for the psychological sense of community. They are not just people with broken hips, impaired hearts, or strokes.

Chapter 6
Planning Programs: Social Security and Head Start

I shall begin with an example that was seared in my mind when I was 13 years old. Franklin Roosevelt had just taken office. In the campaign he had advocated conservative practices. By the time he was inaugurated he realized that he was facing an economic catastrophe. One of his first acts was to close all banks because they did not have adequate reserves to pay their depositors who already were withdrawing their money.

I attended school across the street from the bank. When I went home for lunch—no school cafeterias in those days—I saw crowds of people huddled before the doors of the bank on which an announcement of the closing was posted. The people stood silently, grim, looking at each other anxiously.

The reader may have seen the excellent documentaries on The Great Depression on public TV. To someone like me who lived through the depression and was together with my parents and siblings forever scarred by it, the documentary inevitably looks pale compared to the personal memories it engenders in me. Words are signifiers but not a substitute for personal fears, puzzlement, and helplessness.

America always had severe economic depressions. But up until the Great Depression only small radical groups argued that governments had the obligation to come to the rescue of those citizens least able to sustain themselves without jobs or external help. At best, they received sympathy, with which they were supposed to eat or clothe their children or pay the rent. Government had no moral obligation to come to their help. To the modern reader that will sound callous but it was not meant to be. In those days of rugged individualism citizens were, so to speak, on their own in bad or good times. And even if the bad times were not of their making, it was the luck of the draw and one's mettle was tested by how well they overcame their adversity. All this was occurring as waves of immigrants learned that America's streets were not only not paved with gold but rather contained potholes they never imagined existed. If anything is clear from the proceedings of the American Constitutional Convention, it is that citizens should oppose and fear intrusive government in the lives of individuals and families. Such intrusions are steps to tyrannical government. In the days when Darwin was not yet a household name, the American conception of the survival of the fittest was a world of natural and man-made catastrophes, e.g., the disasters of old age. That view was explicit in the saying that "government is best which does the least."

S.B. Sarason, *Centers for Ending*, Caregiving: Research, Practice, Policy,
DOI 10.1007/978-1-4419-5725-2_6, © Springer Science+Business Media, LLC 2011

Let us fast forward from 1932 to 1935 when the social security legislation was enacted. (If memory serves me right, the legislation passed in the senate by one vote.) It is hard to exaggerate how big a break was with moral traditions that legislation represented. Over their working lifetime employees and employers would contribute to a federal administered fund that would pay the employees a monthly amount for as long as they lived!

There were several reasons the legislation was passed but I shall briefly list only some of them.

1. Twenty percent of the working population was unemployed and it was the older cohorts whose future ranged from grim to hopeless.
2. State and local governments had little or no resources to cope with a depression that was getting worse. Soup kitchens were no answers, people were fearful and restive, and more than a few presumably knowledgeable people proclaimed the end of free market capitalism. As one pundit put it, "Mussolini got the Italian railroads to run on time. What we need is someone like him to put our economy back on the rails."
3. FDR had been in the New York state legislature and then the governor of the state. When he was governor, he surrounded himself with individuals who knew firsthand what was happening to millions of Americans in New York. What he began to see during his New York years was that the massive economic depression posed a challenge to conventional political morality insensitive to the fact that millions of people were being robbed of the opportunity to be productive members of society, to be able to take care of themselves and their families. Steinbeck's novel *The Grapes of Wrath* described and posed the moral issue which could not be glossed over. When later in his presidency FDR spoke of a nation in which one-third of the people were "ill fed, ill housed, and ill clothed," the political and the moral issues were forever wedded.
4. I shall not attempt to explain why and how he was sensitive to the sufferings of people who for reasons not of their making had to radically change their view of themselves and their future. When he contracted polio, his desired political career seemed ended. That he suffered the "slings and arrows of outrageous fortune" can be taken for granted, as we also can assume that he knew full well that young children more than any other age group were by far victims of the disease. He knew what that had to mean in the minds of its victims. It is not happenstance that he started the March of Dimes Foundation primarily to render service and help to those victims. He took delight and pleasure to go to Warm Springs in Georgia and swim in the pool with young polio victims. And we are justified in assuming that he knew he had the financial resources to receive the best of care, a set of circumstances not available to most people, young or old. In fact, at the time the social security legislation was before Congress he considered legislation for a National Health Insurance program. But he was advised that to propose such a program might well result in the defeat of both programs. What I am emphasizing here is that for FDR those were times when government had to act for both moral and economic reasons. And what made his "New

Deal" new was taking action for moral reasons. I should mention here that he appointed Frances Perkins as Secretary of Labor, the first woman to serve in a cabinet position. And he did that for both moral and political reasons.

5. Because of his crippled condition traveling around the country to see and hear firsthand what people were thinking and feeling was not possible. In this regard he was fortunate to have a wife, Eleanor Roosevelt, who was, in his words, his eyes and ears. She was one of the most remarkable women in American history. She may not have looked it but she was no shrinking violet. She traveled all over the country, talked to, and met with countless individuals and groups, insisting on one occasion to be taken deep in a coal mine to see and judge the working conditions of miners. And she wrote mountains of memos to her husband reminding him of his moral obligation to help those who could not help themselves. He spent most of his time in the oval office but his "eyes and ears" were all scanning the country.

What does all of the above have to do with facilities serving aged people today? In light of what I have presented in previous chapters, my answer should not surprise the reader. If the director of these facilities does not seriously feel obliged to get to know how the residents experience and judge the care they are receiving, he may be the top executive, but he is also the top bureaucrat. Between him and the residents are layers of administrators and caretakers, but if the executive relies solely or primarily on what they say is going on, he is shortchanging himself and the residents. Similarly, if people in the political system with responsibility for programs and legislation concerning old people (a) have no first-hand knowledge of the differences between facilities for the have and have-not, (b) and if he has little or no understanding that the label "aged" stands for a very heterogeneous collection of people, (c) and if he has not struggled with an answer to the question of what we owe old people who are dependent, sick, and frail and with little or no resources, (d) and if he is not facing up to the fact that year-by-year the number of aged people will dramatically increase, that politician is part of the problem and not of the solution.

Before going on to other examples there are several things the reader should keep in mind in reacting to what I discuss. First, in the realm of human affairs major social problems are not capable of "solution" in the sense that four divided by two is a solution. Second, for these problems, and because there are many problems, resources are always limited, and choices in kind and degree are inevitable. Third, there is more than one criterion for making choices and compromises but one criterion should be the moral one. If that is not in the picture, it is inexcusable. Fourth, the planning process is not linear, totally objective or rational. Not in the world of earthlings. Finally, in the arena of human services it is not enough to have good intentions, it is not enough to have decision-making power, it is not enough to judge how those you serve "appear" to think and feel. Unless you make a serious effort to help those you serve to feel safe with you, to trust you, to reveal themselves to you, you are confusing the world of appearances with personal realities. We are not born programmed to make such an effort and I cannot give a persuasive explanation for why most people cannot or do not make such a serious attempt.

To say we are imperfect beings is to state the problem, not to clarify it. In clinical fields such as psychiatry, psychology, and social work, practitioners are warned of over-identification with clients because it interferes with objectivity. But that begs the questions: How do you learn when the fine line between productive identification and unproductive over-identification has been crossed? I regard what these physicians say as a copout. Nobody is suggesting that physicians should allow themselves to become emotional basket cases. And nobody is suggesting that they become stone-faced, cold, uninvolved, or disinterested clinicians. But one can suggest that we study those physicians, albeit a minority, who are superb clinicians and whose interpersonal style radiates sincere interest and warmth. Why are they a minority? If you have no selection criteria relevant to caring, compassion, and personal style, why on earth should you be surprised that you end up selecting a population skewed in ways contrary to the goals your rhetoric purports to achieve? In the arena of the human services—I include education here—the selection process for all those who will be helpers in one or another way has at best an unknown validity and at worst demonstrable invalidity. In regard to programs for aged people, my observations over the years and my experiences in such programs are a very negative one.

The reader may be surprised that the next section is about the Head Start program initiated in the mid-1960s for children 3–5 years of age living in the ghettos of cities and who will attend the public schools where the level of school performance has been scandalously poor. As I shall indicate, crucial issues about the criteria for selecting caretakers–teachers of these disadvantaged preschoolers were flawed in precisely the same way it has been for those who will work with the cohort of the most aged people. Needless to say, these programs have taken place at different times and in social-cultural, political contexts. But in one respect they are very similar: issues about the criteria for selecting caretakers were at best grossly superficial and at worst self-defeating of their purposes.

When Head Start was enacted I said in a public lecture that it could not and would not achieve its laudable goals, although I also said that if I had been a member of Congress, I would have felt morally obliged to vote in favor of it. Let me explain why.

Head Start

The social security act was a form of insurance that an individual joining the work force was guaranteed that at retirement he or she would receive a monthly check the size of which was determined by years of working, amount earned, and the contributions of the employer and employee. The rules were cut and dried and could only be changed by an act of Congress. It made no difference who was administering the fund. The rules for post-retirement payments were unambiguous and impersonal. It was a form of contract reflecting what working people had a right to expect after a lifetime of work.

Head Start was neither a contract nor a form of insurance or a guarantee of predictable outcomes. It was primarily a moral response to the consequences of slavery, the 1954 desegregation decision, and the inequalities and barriers that the decision exposed about the preschool experiences of black children. Where it was similar to the social security legislation was that President Lyndon Johnson seized the opportunity, given the social-political climate of the times, to take action. It is not irrelevant that his hero was President Franklin Roosevelt. It is also relevant to note Johnson had been a teacher in a school in rural Texas where the negative correlation between poverty and school performance was brutally clear.

Why is there no credible evidence that children in Head Start do discernibly better in school than children who have not been in Head Start? That, I hasten to say, does not necessarily mean that an undetermined number of Head Start children did not benefit in one or another way from their participation. We simply do not know. What we do know is that if their school performance improved, it was at best very small. There are several reasons.

The first reason is based on the implications of an irrefutable fact: What the children will experience and learn will depend on those who teach them, those who have a direct relationship to them. A context of learning is quintessentially one of interpersonal relationships—teachers of students and students with students, the latter fostered and supervised by the teacher. What does a teacher need to know, reflect, and implement if the context of learning is to be productive of goals? Has the person read a book, or attended some lectures, or brief workshop, or had been a babysitter, or caretaker of young siblings, or had experience only with individual children but not with groups, etc.? I ask these questions because the administrators of Head Start in Washington glossed over or ignored such questions, and those in charge of a Head Start program in local cities hardly asked them, *if only because teachers and their classroom aides were going to be paid a scandalously low salary, thus insuring that inexperienced, poorly educated teachers and aides would make up the pool of applicants.* (I remind the reader that what I am describing is identical for aides in nursing homes.)

The problem was exacerbated in the case of Head Start programs in our urban centers where militant blacks were opposed to hiring white people either as teachers and aides. Several members of the Yale Psycho-Educational Clinic offered clinic services to local Head Start programs but in so many words it was conveyed, sometimes openly, that blacks should teach blacks. Our services were rejected. That happened elsewhere around the country.

The second reason for Head Start's weak effects also involves the quality of the "caretakers"; in this case I am not referring to Head Start personnel but to teachers in the schools the Head Start children would go. In the realm of the human services it should be axiomatic that for any major problem you quickly examine who the caretakers are, whether they are in nursing homes, hospitals, or schools. What I shall now discuss briefly I have elaborated in publications beginning in 1965 (Sarason, 1969, 1971, 2004).

Implicit in the rhetoric of proponents for Head Start is what I have called the medical conception of contagion. Concretely, Head Start was viewed as preventing poor minority children from catching the "disease" of lack of motivation for and disinterest in learning when they entered schools. Nobody in the political establishment as well as in the intellectual community could say that out loud for fear of alienating the educational community whose political support was vital. In fact, Head Start began about the same time the indices of educational outcomes steadily began to go down. Head Start was intended to inoculate its preschoolers against the "diseases" of disinterest and dislike of boring, unstimulating classrooms. Head Start was not the educational equivalent of the Salk vaccine! I should point out that the educational establishment wanted Head Start to be in or administered by the public schools, a suggestion to which the black community gave a militant, negative "no."

Earlier I discussed the work of the poet Kenneth Koch in a Spanish Harlem elementary school and in a dispiriting nursing home in Manhattan's lower East Side. Before Koch arrived at these two sites the children and the aged people were unmotivated, unchallenged intellectually and artistically. They became alive. As another poet once said, education is not filling empty vessels, it is lighting fires. Koch lit fires. On a larger and longer scale Levine (2001) and Bensman (2000) have written books of the work of Littky and Worshor, and Deborah Meier with at risk, poor, minority high school students. And when you read the results they present, it is refreshingly inspiring.

Nursing homes and schools evoke very different imagery and purposes. To compare them would seem to be a case of mixing apples and oranges. However, let us not forget they both are fruits whose appearance, taste, and nutrition depend on who takes care of them.

Why is it that in the human services we do not seriously and routinely ask: Who should be a caretaker of this or that needy group? What do they have to know? How can this knowledge be honed by training? Do they ignore such questions out of a lack of sensitivity or out of fear of the answers?

I started this chapter with the Social Security Act for several reasons. The first was that it was the first time in our national history that all people in their older years could expect monthly payments regardless of whether they were in the minority who would not need it. The arrangement was straight forward and impersonal, as insurance plans usually are.

The second reason was to contrast the act to what became apparent two decades later when the size of a coming baby boom could not be dealt with by towns and cities, especially the latter, which, because of the Great Depression and the exigencies of the war years, had caused an inadequate educational system to go steadily downhill. For several years the federal government did nothing because of its longstanding tradition in matters of education. But in the early 1950s the government decided to give financial aid to cities to bolster the quality of their programs for schools in poverty areas. It was expected that such support would cease in 2 or 3 years when cities would be able on their own to get "over the hump." What went ignored or totally unnoticed was defining the problem as a financial one, thus missing the trees for the forest. That is to say, schools have been inadequate for decades and whatever reform efforts that had been tried were ineffective. What happened? With each succeeding year the government has steadily allotted increased funding to reforms which have proved ineffective.

There is one thing all of the reform efforts have in common: they seek to change the behavioral, interpersonal regularities in the classroom and thereby improve educational outcomes. If the social security program achieved its goal, it was because its goal was *not* to change the behavior of and interactions between people. *But changing interactions between people is precisely the goal in the classroom and institutions for aged people and that is why in earlier chapters I used my experience with and in nursing homes and hospitals.* I wanted to alert the reader to take the obvious seriously: Whether you are observing an individual, an interaction, a classroom, or a nursing home, or a family, you should not assume that the world of appearances is a reflection of the world of inner reality.

My fear is that if and when they are seriously planning for the millions of aged people who will want or need a center for endings, the planners will ineffectively seek to build more of what we already have, and by "home" they will mean physical structures. We certainly will need these structures, just as the baby-boom generations after World War II were a financial bonanza for architects and builders. If the schools were new, the purposes, rationale, quality of classroom interactions remained what it has long and unproductively been. Educational planners put old wine in new bottles. That is what I fear will happen in a decade or so when it will no longer be possible to ignore the need for new facilities for the aged people who comprise the cohorts of the post World War II baby boomers. I fear that the shortcomings of Head Start and educational reform will be evident also in whatever planning will be done in regard to aged people.

Chapter 7
The Haves and the Have-Nots

The major purpose of this book was to ask and discuss one question: Why has our society not seriously confronted the fact that in a few decades we will have for the first time the largest number of aged people, the consequences of which will transform our society? How it is now transforming the society is cause for concern and predicting the course of those transformations over the coming decades will depend in part on when the powers that be wake up to what is coming down the road. When in 2005 hurricane Katrina decimated the Gulf States, it was known that Katrina was on the way but we were unprepared for the strength and reach of a hurricane 5, a strength heretofore unknown on this continent. The reverberations have been political, racial, and economic, and there is good reason to expect that they will continue both in predictable and unpredictable ways.

In my case it was inevitable that as a result of the fact that I am a very old man, my experiences in my later years would remind me how often it has been that the society's plans for its major social problems have been belated and far from achieving desired goals. As I write these words in 2007, Washington and the larger society are caught up in and conflicted about the substance and direction of new immigration legislation. It would be more correct to say that *finally* the society has come to understand that (a) in a couple of decades our largest state will be dominated politically by Hispanic immigrants or their offspring, (b) the number of illegal immigrants has been and is huge, (c) a large number of states, even excluding those in the southwest, already have many Hispanic immigrants, (d) given the birth rate of these immigrants, what will happen in California will begin to change the national landscape, (e) there is no simple solution to the problems which contain political, economic, racial-ethnic, legal, linguistic, and educational and moral aspects. To put it simply, it is a mess. *But it is not a new mess.* It has been building up for several decades, especially after World War II, and many people in the political arena in California and the southwest knew about it, some with approval, others with disapproval. Certainly the demographers in the federal census bureau provided a factual account of what was happening. As the philosopher Pogo would remark today, "We have met the enemy and it is us." And, I believe, that is what Pogo will say a few decades from now about the way we ignore the fact that we will have millions of aged citizens for whom programs, services, and care will be inadequate, more of a mess than it is today.

S.B. Sarason, *Centers for Ending*, Caregiving: Research, Practice, Policy, DOI 10.1007/978-1-4419-5725-2_7, © Springer Science+Business Media, LLC 2011

Let me now turn to the 1944 legislation, the G.I. Bill of Rights for returning World War II veterans, which many people regard as the most important piece of legislation of the twentieth century because of its transforming effects on American society.

1. The relationship between Japan and the United States had been deteriorating years before Japan bombed Pearl Harbor on December 7, 1941. In fact, the United States had broken the code Japan used and had known that Japan was planning some sort of strike, but never expected it would be Pearl Harbor. That bombing came as a shock because it resulted in the destruction or crippling of our warships there, making the country's mainland vulnerable for attack.
2. The number of deaths and casualties on that day rivaled that caused by the destruction on 9/11 of the World Trade Center.
3. Two things were immediately apparent: It would be a long war, victory would not be certain and Veterans' Hospitals would not be capable by their size and quality of personnel to care for the many thousands of veterans who would need care and rehabilitation.
4. A year after the war in the Pacific erupted, a committee was given the task of determining—really an informed guess at that time—how seriously the shortages were of the mental health and medical personnel needed to care for the scores of thousands of veterans who directly or indirectly would be casualties of the war and many of whom would require care for many years after the war. When the war began the United States had a population of one hundred and fifty million people. At the war's end somewhere between fifteen and sixteen million people had served in the armed services. Today the population is three hundred million people.

   If you assume as a minimum that ten percent of the millions of aged people will be in the population three or four decades from now and will need special facilities, programs, and care, which many will be unable to afford, the reader may better understand why I have written this book and see the future as a gathering storm for which the government is as prepared as it was for Katrina.
5. I do not know who was the major influence in choosing members of the 1942 committee. Whoever he was he chose a group who can best be described as "Young Turks" capable of thinking outside the box and dissatisfied (too weak a word) with the deplorable quality of the personnel and services of the hospitals administered by the Veteran's Administration. One of the first decisions they made was that if veterans were to get the best possible care, as they deserved, a totally new and bold system would have to be created which would require that new hospitals and clinics be built as near as possible to a university and hospitals where clinical services were informed by and based on research. And the relation between these centers and the new hospitals would be *formally* institutionalized so that training of all new personnel would involve clinical training in university and VA hospitals.

Personnel in the new hospitals would be of a quality justifying appointment in the medical school. The decision meant that a new VA would be created. It is hard to exaggerate how revolutionary that decision was. It is also hard to exaggerate how costly new systems would be but if the society wanted the best for those who were casualties of war, so be it, it was a price morally justified. Veterans who heretofore needed services but who did not have resources to access the quality of service in the private sector would now have access to new university—VA hospitals and clinics.

The VA has had its ups and downs. For 17 years I was the director of Yale's graduate program in clinical psychology and a consultant to several VA hospitals. If I had to identify one source of the quantity and quality of service and the lowering of staff morale, it would be that VA hospitals were under the jurisdiction of a central office in Washington which in turn was embedded in the executive branch of government which in turn had to deal with Congress, lobbyists, and a variety of passionate pressure groups. The simplest way to describe the inevitability of ups and downs is to say that this layering of power and responsibility means that as you go "up" from one layer to the next one, to the next one, etc., decisions are made by people with no firsthand experience of the complicated culture of a modern hospital, and that is true in the case of the president who is expected to read, study, and understand a budget of fifteen thousand pages. Yet despite this the VA has done a very credible job, and for one reason only: Whereas before World War II the VA was a passive, conforming system with essentially no supporting constituencies even with a semblance of power, after world War II the VA had one powerful, articulate constituency—the university and its medical schools. Each alone or in combination knows how to organize support of other well-organized groups.

In 1944 the landmark legislation called the G.I. Bill of Rights was enacted. It is hard to say if the 1942 report in any way led to or influenced the boldness of the G.I. Bill of Rights. As I indicated in earlier pages, the G.I. Bill made it financially possible for any veteran who wanted to go to college or the university to pursue his or her goals. The veteran had only one barrier to overcome: admission to a regionally approved educational institution.

There was some opposition to the Bill, especially from Senator Bilbo of Mississippi who never missed an opportunity to proclaim publicly the most demeaning things about blacks. The thought that a black veteran would take advantage of the Bill to further his education, especially if he wanted to go to a white college in Mississippi, sent Bilbo up a wall. And there were well-known leaders in higher education who feared that allowing all veterans to take advantage of the Bill would lower academic standards. The Bill passed because it was expected that only about two hundred and fifty thousand veterans would apply for an educational program. In fact, several million took advantage of the Bill.

How do we explain the prediction that only a paltry number of veterans would take advantage of the opportunity? What does that suggest about how we form opinions of what people are and what they will or can become when we have made no effort to give a person an opportunity to show that they may be wrong?

If a half century ago we asked people if women were capable of becoming scientists, heads of major corporations, governors, members of Congress, Supreme Court justices, and flying airplanes in battle, practically no one would have nodded assent.

But when events converged to make it possible for women to express their worth and to take actions to pursue them, conventional wisdom was exposed as being neither conventional nor wise. That is what Kenneth Koch exposed in creating the conditions where ghetto children in a Spanish Harlem elementary school wanted to write poetry. And he did it again with aged residents in a nursing home. And it was also done by Schaefer-Simmern who created conditions where institutionalized, mentally retarded individuals engaged in artistic activity (Schaefer-Simmern, 1948).

The G.I. Bill offered the returning veterans an opportunity for new experience. It was expected that few would take advantage of it. They could not have been more wrong. Generalizations about veterans were misleading and demeaning.

In both World War I and II a major problem in the armed forces was illiteracy. The army could have (and probably did) give essentially illiterate people an honorable discharge. In their book *The Uneducated*, Ginzberg and Bray (1953) did an exhaustive study of illiteracy in the armed services, a project given the support of General Eisenhower who was then president of Columbia University. In that book they describe an experiment dreamed up by some officer in an army base. Several small groups of these illiterates were given a choice. They would be given an honorable discharge or they would live together with a teacher for no more than 120 days to achieve a reading level at least on the fourth grade level. Those who met the criterion could remain in the army, those who did not would receive an honorable discharge. They met the criterion in 92 days. When they were later asked how they felt about their experience, they could not have been more gratified. Most of them said that when the war was over they would seek more education. They had acquired a basis for wanting to learn more.

Wars change everyone and everything and World War II was certainly no exception. The offspring of the World War II veterans and non-veterans will have had on average far more formal education than their parents. They also will have experienced a generational gulf between their parents and themselves. They had a different view of the future and what the goals of living were. They had greater expectations that they should and could experience "the good life" as they defined it in a society that was undergoing fast cultural change; probably the fastest in our national history. There were (are) three features of that change relevant for the issues I have raised in this book. The obvious one is that the role of women and their self-attitudes changed dramatically. The second was that marriage "as we knew it" accompanied the first change. Half of all new marriages ended in divorce, women postponed having children, and it was not a source of comment when couples did not see the point of making their relationship legal. The third feature was less obvious because you could not "see" it. The economic and educational gulf between the haves and have-nots slowly but steadily began to widen.

How will these and related changes manifest themselves in the course of the next few decades when we will have millions of old people who on average will probably live longer than old people today as a result of medical research?

On what grounds can one deny that we have no reason to begin now seriously to try to answer the questions? You cannot answer questions you do not ask. It is only in the last decade that the general public has become aware that the earth's temperature has increased and that in scientific circles there is a controversy about why it has increased. Is it because of our ever-increasing use of fossil fuels which are not a limitless resource? Or are we entering a geological age which could last a thousand years or more? Or is it one of nature's flukes which happen every now and then for reasons we do not understand so let us not worry?

I have listened to many lectures and symposia on global warming and although all explanations had their passionate adherents, no one ever said or hinted that we were not faced with a serious problem about which we should do little or nothing. As one scientist said, "It makes no difference which explanation you accept, we cannot as a society sit on our butts and do nothing." Even scientists who do not buy the fossil fuel explanation said that there were many reasons why we should reduce our dependency on oil. No one I heard offered any concrete steps on the level of action.

If there is controversy about whether global warming is, so to speak, "real," there is none about the fact that we will over the next few decades have millions of aged people, a fact that could, or may , or will transform American society. We cannot ignore or gloss over that fact. But we have and still are keeping our heads in the sand. I am amazed how many people who are 50 years or more have little knowledge of the crazy quilt features of programs, services, and facilities for aged people and how they vary in size, cost, quality, and the hoops some aged people have to jump through to get what they hoped to get, but usually are disappointed. Let me give just a few examples.

Each of a married couple had worked for years and when illness forced one of them to retire the other also retired. Their combined income when they retired was $55,000 a year. They had two children who were able to get a state college education through student loans, plus support from their parents. The family lived in a house they owned and recently made the final payment on their mortgage. After federal, state, and local taxes and the usual expenses for clothing their children, upkeep of their house and their car, food, movies, etc., they could save little. In addition, husband and wife had their monthly or weekly social security payments deducted and the husband also had deducted from his paycheck his contribution for health insurance in his employer's health plan.

John was an old widower whose wife had died 2 years ago from cancer. He still lives on the top floor of an old, seedy three-story house in a rundown neighborhood. His wife had always been sickly and could not work. John worked as a janitor in the school system. His pension was about six thousand dollars a year and until his wife died their combined social security checks were nineteen hundred dollars. Neither of his two sons graduated high school and they had low-paying jobs. John and his wife had few friends who now were either dead or moved in with one of their children. John had always been proud that he had never applied for public assistance. But 2 months before his wife died he went to the public welfare department for help. He was given food stamps and was told that he would be visited by a social worker who could determine if they were eligible for other payments and/or services. When the next week the social worker arrived, she saw that John was unsteady on his feet. When she asked him how often he walked up the three flights of stairs, his answer was about four times a week. When she went back to her office (there was no phone in the apartment) she called the doctor in the clinic of a local hospital and was told by him what she had already concluded: John's wife would probably not be alive more than a few weeks and that it was a wonder that John had in the past year fallen down the stairs only twice. Could he, she asked, get John's wife into hospice? He said that he doubted they had an empty bed but they could send a hospice home aide to help in the care of his wife. The social worker then began to determine where she could place John who could not or should not be living alone. Could she persuade him to go to a nursing home until a suitable living arrangement could be found? It was a question she asked several times a day. At this point, John's and his wife's problems were less a matter of money and more a matter of limited, suitable placements.

The first example above is about aged people who are not poor but conventionally would be considered lower middle class, as most baby boomers would be. If they seek to enter a total care facility, they could not afford the entry fee and the monthly charge. The state does not build such facilities; they are built by corporations in the private sector and who understandably seek to be profitable. Even if their modest home could be sold for $150,000, they would be unable to live in such a facility for more than 2 or 3 years, a fact that will terminate any conversation with a representative of such a facility. What if the example was about a couple in which the wife needed to be placed permanently in a nursing home and the husband would sell his house and rent or buy a small apartment or condo? He would understandably want what by his standards would be a "nice" apartment just as he would want a nice nursing home for his wife. It would not be many years before his financial resources would be insufficient. Or take the example where the wife goes to a nursing home and he goes to an assisted living facility. In 3 years at most their financial resources would be insufficient. There are other scenarios about how lower middle class people would have a very problematic future financially and psychologically.

The "faces" of poverty are many, bewilderingly so. The struggle with a precarious existence is nothing new to poor people. Many of them have received public assistance of some kind. But they maintain the belief that they are independent, capable of making choices among whatever options they may have; and the number of options is small. If they do not value themselves highly, they still resent those in the larger society who do not accord them respect or recognition as individuals who did not will their low status. They had fantasies about a better future but by early in mid-life those fantasies have vanished. Life does not get better for them as they become more subject to illness and disease and become more and more dependent for support from government agencies.

What happens when a poor, aged person decides, or someone then decides for him, that it is not safe or wise for him or her to continue to live alone and should go to a facility which will oversee his care? Let us assume that he agrees with the suggestion. In which kind of facility will he be placed? Where is it? What choices does he have? The fact is that because he exists on public welfare for financial support, he will have little or no choice. His social worker will tell him there is a shortage of beds in the state or city, there are waiting lists, she will do her best to get him placed as soon as possible, and she cannot promise that it will be near where he has lived. The social worker knows which facilities are better than others, that even the so-called acceptable ones are not so acceptable; she would not like to be in any of them. She wants more than the feeling of being safe. On what basis can we deny that "more" to the aged poor person? Why probe when what you are told you will consider unattainable and unrealistic? Who in these facilities have the knowledge and willingness to create the conditions in which the "more" stands a chance of being manifest? Kenneth Koch did it via poetry with ghetto students in Spanish Harlem, and with aged poor people in a seedy nursing home. Schaefer-Simmern did it with long institutionalized mentally retarded individuals via visual art.

I did not write this chapter to alert readers to what they already know. We have many haves and a far greater number of have-nots. My purpose was by way of raising a moral question: What as a society do we owe aged people who near the end of their lives will need or want to be in facilities in which their needs are met and where they will experience care and compassion as well as opportunities to engage in activities which contribute to their sense of worthiness? Put it this way: *What is the opposite of warehousing people?*

That was the question asked and answered by the new Veterans Administration after World War II. It is hard to exaggerate the differences between the old and new VA. The difference is not comprehensible unless one knows that the old and new VA were answering the "What do we owe them" question in radically different ways. And was not and is not the "owe" question the central point in the history of women and blacks? And I feel justified in asserting that the "owe" question in regard to aged haves and have-nots, whose numbers will dramatically increase and for whom the shortage of available facilities will also dramatically decrease, will remain unanswered. Those are brute facts, not exaggerations of a fevered imagination.

I am not a seer. I will not offer any specific predictions of what may happen if we continue to ignore the issues and facts I have discussed. I am sure that readers of these pages will have been asking themselves where will the money come from to pay for whatever costly plan is adopted. It is a legitimate concern that brings to the fore again the moral basis by which as a society we make decisions about how to allocate limited resources. We never have and never will have sufficient resources to support all programs to the level we desire. We know ahead of time we will fall short of the mark.

It is hard for me to envision that serious planning for coping with the problems I have discussed without the leadership of a president who has not only curiosity and courage but also knowledge and wisdom implicit in the sense of morality in the "what do we owe aged people" question. That is why in the next chapter I elaborate on issues of political leadership.

Chapter 8
The Need for a Presidential Commission: Some Caveats

In what follows I am making the assumption that in regard to the care of aged people we have a major problem which we must confront, it is not one of the proportions and complexity of which we can continue to ignore. Readers who do not accept that assumption need not waste their time reading this chapter. The assumption leads to the suggestion that the president appoint a commission charged (a) with reviewing and evaluating the present programs and services for aged people who need them, and (b) with making recommendations for changing and improving them, and (c) with making their findings available in 2 years.

Choosing the Members of the Commission

Let me start with the commission President Reagan appointed to probe why the educational outcomes of schooling were depressing despite all efforts and a lot of money to improve them. As is usual he appointed a well-known public figure, Bart Giamatti, president of Yale, to the commission. It was likely that the president did not care if Giamatti was a Republican or a Democrat. But we can safely assume that he was told that Bart Giamatti took a very dim view of schools of education, of the poor quality of too many school personnel, and of the low standards which students were expected to meet. I assume the two presidents met and that President Reagan was more than impressed with the fact that Giamatti was brilliant, a man of very strong opinions, and truly believed that the inadequacy of our schools put "A Nation at Risk," the title of the commission report. I have little doubt that when they met, they took delight in the fact that both loved baseball; Reagan had been a radio announcer for baseball games and Giamatti had publicly said many times that the one presidency he had always longed for was that of the American baseball league in which his beloved Boston Red Sox played.

I start with Giamatti because he was the only member of the commission I knew. I did not know him well but I heard talks he gave and what his views on schools were. But I knew one other thing: Bart Giamatti had no experience in classrooms as teacher or observer, he was very poorly read in the educational literature, and spent his years at Yale which once had a graduate department of education which had been eliminated by a previous president who held the same views of education

S.B. Sarason, *Centers for Ending*, Caregiving: Research, Practice, Policy,
DOI 10.1007/978-1-4419-5725-2_8, © Springer Science+Business Media, LLC 2011

as Giamatti. To say that Yale has always been hostile to matters educational is something of an understatement.

Was the level of expertise of Giamatti exceeded by other members of the commission? Excluding Giamatti there were three other university presidents. I make the assumption that anyone knowledgeable about the duties and obligation of a university president will be surprised if anyone who holds that office knows or has direct experience in schools.

There were five members who were either a state commissioner of education, or on school boards (local or national), or superintendents of schools. I have known many of school board members and they tend to be political appointees who over the span of their tenure have learned a lot about budgets; renovating old schools and building new ones; and the adversarial relation between schools and teacher's unions, between state boards of education and schools, and between schools and federal directives. As for superintendents, their knowledge of what is going on in the classrooms and the system as a whole depends on what lower layers of administration tell them. I need not discuss the few remaining members of the commission: two principals and three university academics. It would not alter my conclusion that it was predictable that the final report would be a collection of generalizations which lacked the substance and concreteness of recommendations.

I cannot refrain from noting that the commission report said not a word about the steady, dramatic increase in the number of students who were being placed in special programs for the mentally, emotionally, and physically handicapped individuals, programs mandated by the federal government. Almost immediately after passage of the legislation for these programs, serious and concrete problems arose, finances not being the least. Not a word was said in the report about what was already happening. And nothing was said about the low quality of educational research.

Why the Reagan commission died a quick and quiet death will become clearer in the next section which will prepare the reader for the complexities that arise when it comes to choosing the members of a commission on needs and programs for aged people, assuming such a commission is formed.

The 9/11 Commission

The 9/11 Commission focused on all personnel at all levels in all agencies with responsibility for national security. Why was the administration so utterly unprepared for the destruction of the World Trade Towers in Manhattan? Why were they unable to "connect the dots" in the mass of intelligence data that had been collected? Part of the answer was already public knowledge before the commission was appointed. It seemed as if different intelligence agencies would not reveal what they knew to each other. President Bush had been apposed to a commission but as a result of public pressure, especially from families who lost friend and kin, he gave in. He knew that he had to appoint individuals who had first-hand experience in the ways of the Washington scene. By that criterion he appointed a bipartisan commission. They did a superb job of investigation and made concrete, clear conclusions. Their report contains few generalizations and distractions. They recited line and verse for what they ultimately proposed. Testimony to their knowledge of the Washington political scene is that after the report was made public they devoted themselves to bringing the report to the country at large. They did not write a report, go home, and that was that. They did not want this report to languish on a shelf in the government archives. They knew the game and the score of the ways of the Washington scene. The consequences of the report were bold and many. Their report and that of President Reagan's "A Nation at Risk" could not be more different from each other. "A Nation at Risk" managed to be dry, dull, and uninformative and unconvincing that a nation was at risk, a risk I have been writing about for decades. I like to quip that I am opposed to sex education in schools because they would make it uninteresting. "A Nation at Risk" managed to cause its readers to wonder what the risk was.

A Commission on the Aged: What Are the Problems?

The title of this section is by way of indicating that society generally and government in particular have not been prepared to think seriously about what the future certainly holds in store, and the problems that are already evident today. Political leaders have exercised no leadership and in the halls of congress no one has issued a wake-up call. In newspaper and other mass media little has been said to alert, educate, and arouse people about what may or will happen. It is understandable, is also self-defeatingly shortsighted in the extreme, and is no excuse for the absence of leadership to take the initial step of appointing a commission. Since I will not be alive when the society will be compelled to recognize the obvious, I felt compelled to write this book for a president who will be forced to appoint a commission. And when that happens the danger will be that the charge to a commission will not ask for answers to some concrete problems. There is, however, a prior question which is as bedeviling as it is crucial for everything that the commission will or will not do. And that question is why I discussed and contrasted Reagan's and Bush's appointees.

Criteria for Appointment of Commission Members

Physicians, nurses, physical therapists, social workers—these are personnel who have direct experience with and responsibility for aged people in need of some form of assistance in one or another facility or program whose assessment is required as a basis for placement and programming. For example, when I applied to be a resident in a total care facility, my personal physician had to fill out a form about my physical health, infirmities, and ability to live independently. That meant that once I was in my own apartment I would direct all health questions to my personal physician. However, when as a result of a fall I had to be placed in the nursing home, all matters of health became the responsibility of a geriatric physician who spent two half days a week in the nursing home. I never saw or met her, she never examined me. After 3 weeks I demanded to see her to register my complaints and questions about the quality of the services I was receiving. To my surprise she agreed with me! There was little she could do, given her limited time. That infuriated but did not surprise me. Why was I not surprised? Let me briefly explain because it is relevant to the selection of members of a commission.

From 1942 to 1960 I came to know a variety of medical specialists, either as colleague, or friend, or observer. In all instances I was primed to judge them by previous experience with families whose physicians had strongly persuaded them to institutionalize their mentally retarded child, a degree of persuasion that overcame parental reluctance to accept their advice. The advice was well intentioned but it was misguided, insensitive, and caused sustained guilt and anguish. In a book *Psychological Problems in Mental Deficiency* (1949) there is a long chapter about the unwillingness or inability of physicians to listen to what parents are saying and implying. That explains why in subsequent years it should not be surprising that I judged all my medical friends and colleagues as I ultimately did and still do. Less than a handful of them received a positive judgment from me. They, of course, do not see themselves as I did. Yes, they were all more than just competent as diagnosticians of the body. All else was patient "noise."

Beginning in the 1960s physicians, for the first time, were subject to public criticism for their lack of caring and compassion. Many medical schools made efforts to be responsive to these criticisms but the steps they took were token gestures. There is no evidence whatever that these gestures had any desired consequence.

Why did I start this with the physicians? Because only in recent decades has the field of geriatric medicine taken shape and in appointing commission members, the president would probably be advised to appoint a geriatric physician as chairman and in addition a couple more medical specialists. That would be a huge mistake because, however, expert in physical diagnosis and choosing the appropriate medications, their understanding of the psychological substance and dynamics of the phenomenology of the final years of life is not impressive. And of the few among them who do understand, they soon learn that the role of the whistleblower comes with a price. What the president must do is to locate these few and give them a forum to which their views can be presented. The people we do not need on the commission are those who will say we have been on the right roads, we know the problems ahead, and what we will need are the resources to do what we have to do. Ironically, that will be the kiss of death in an arena in which people know all too well that death is not too far off. Taking care of bodies is one thing, taking care of minds is quite another thing.

Let me conclude this section by turning to the one group who spends the most time with sick, frail, dependent aged people: the aides, who are the least educated, paid the least, get no training, and who are employed more because they are available and willing to do the work assigned to them. To say they get no training is not an overstatement. *It may come as a surprise to readers when I say that I have never met any physician, geriatric, or otherwise, who disagreed with what I have just said, and ditto for nurses.* And, yet, they are fearful to go public. They feel they are locked into a system over which they have no control or influence.

When I wrote this section and asked a physician friend to read it, he asked me, "If the president asked you to recommend a chairman for the commission, does no one come to mind?" I blurted out totally, and I mean totally, honest answer. I said Dr. Morris Wessel, a clinical professor of pediatrics at Yale. Why Dr. Morris Wessel has always been in private practice? He is not a researcher, but he has made known his criticisms of pediatric training and practice in numerous articles. He is nationally known and respected for the stormy petrel he can be. He has been accorded many honors. How you think about and treat sick children, their families, deaths in the family, the hospital culture, professional insensitivity, and arrogance—these are the kinds of issues which cause him to express his opinions forthrightly: He was one of a handful of physicians and nurses who created the first hospice in America, a story not yet told, a story about fighting a medical establishment not comfortable with a new entity challenging medical control of the final days of life. Dr. Wessel is not understandable without knowing that he is internally governed by one question: What do we owe sick people and their families? He is not a geriatric physician but is not "only" a pediatrician. The fact is that much that I have said in this book is old hat to him.

Whether they were Republicans or Democrats, the members of the 9/11 Commission knew the Washington scene extraordinarily well, a fact that made the obligations of party affiliation an irrelevancy. Dr. Wessel knows the medical culture and scene extraordinarily well. The only negative about him is that he is as old as I am, albeit in far better shape than I am. A younger edition of Wessel I would put on the commission is the surgeon Atul Gawande whom I discussed in Chapter 1. I am sure that there are some administrators who see things as Gawande does, but they do not write.

Aides: The Immediate Pressing Problem

I use the word "immediate" advisedly for two reasons. The first is that aides have the most direct contact with residents in nursing homes regardless of whether they are facilities for the haves or have-nots. Nationally, aides are members of minority groups from this country or foreign ones. To say they are "selected" is true but grossly misleading because they are in short supply, an understatement even if you refer only to the few who are considered "good." There are very few of the few. The second reason is that for all practical purposes they receive no training. Any or all efforts to improve quality of care can only be judged by evidence that aides are implementing appropriately the letter and spirit of the rhetoric of quality care. If tomorrow one doubled the pay of aides, on what grounds should he/she expect an improvement in quality of care?

In the 1950s, and as a result of the rise of militant teacher unions, the society was forced to recognize that the salaries of teachers were scandalously low. As a result, salaries began steadily to increase to the point today where they are respectable. And yet during all those years the quality of the educational outcomes of our schools either went down or remained the same.

Initially at least, the quality of aides will not be improved by money, although that is no justification for what aides currently are paid. What is required initially is a research program on the selection and training of aides. If you know how to select, you have licked 50 percent of the training problems. Select them for what? To make beds, pass out meal trays, etc.? They can do that now. What understanding do you want them to acquire about aged people? How does the understanding help them to see interpersonal relationships in a new light?

I assume that research would not be justified if it only described what aides currently are assigned to do. And I also assume that such a program is being carried out by people who are in agreement with the need to select and train aides in a mold which, if successful, will be discernibly more fruitful for everyone in the facility: physicians, nurses, physical therapists, and, of course, the aged whom they are morally obliged to help.

The question is not why aides are what they are, and to blame them, which is to blame the victim. No one denies there is an "aide problem" but no one thinks about it, let alone writing about it. No one wants to be a whistleblower. I do not know whether the situation is similar in other western countries. It would be interesting and instructive if there are countries where the situation is different. Cross-cultural comparisons have many virtues and chief among them is the possibility of learning that the way we define and deal with a problem is but one way of thinking and dealing with it.

Voices of the Aged

What we think we know about the phenomenology of aged people comes less from the aged themselves than from the diverse professionals who have a relationship to them. Those relationships may be a routine affair for the professional but it is certainly not for the aged person. Who decided that he and/or his spouse (or both together) need to be in a protective environment and are seeking information and advice? For the aged who are poor the problem is not "What should I do?" but "Where will you put us?" And in the not too distant future there will increasingly be middle-class aged people shocked by disparity between what they can afford and what they had expected to afford. I am talking about people experiencing a transition to a new lifestyle, to a strange, puzzling future. Aged people are future oriented and that future is by no means clear. They want to feel and to be regarded by professional gatekeepers as competent, flexible, and pragmatic. Their reservations and fears are kept private.

It may surprise readers that I said aged people are future oriented. The stereotype of the aged person is that he or she reviews and lovingly talks about the past. There is a joke about an aging woman who tells her stockbroker to purchase for her a certain stock. The stockbroker is surprised and says that it may take years for that stock to appreciate in value. She replies by saying she knows that but she wants it for the long term. People laugh at the punch line because it shows how one aged person can deny how numbered her days are, how laughably unrealistic she is. But they will not laugh when I relate when in the morning I take the elevator to go to the exercise room and there are other residents who are going there; and I ask one of them "How are you doing?" I frequently get the answer, "Well, I slept through the night, I woke up in the morning, so I can't kick." That brings smiles and nods to others. They are still alive and the implicit message is, "I want to wake up tomorrow."

In the abstract, professionals know the transitions to the center or the nursing home will not be easy, but in practice they hardly discuss it, let alone bring it up. And what happens when after the transition the aged person is unhappy, or disappointed, or uncooperative, a complainer? The caretakers stigmatize such people with negative labels, recite the rules of the facility, or offer verbal bromides such as "I know it is not easy but give it time and you will see things differently." What it does give the person is confirmation that there is no point in expressing feelings. The zone of what is private increases. It is what a resident said to me, "You can't fight city hall."

Why did I start this section in the way I have? Because a presidential commission will have supporting staff who will go into the field to collect data relevant to the commission's charge. Hearing the voices of the aged is not a matter of ritualistic courtesy. Rather it is necessary to speak directly with the elderly whom we are obliged to help in order to understand their thoughts, experiences, opinions and reactions. I am not recommending anything like a witch hunt. I do recommend that in each facility visited by staff the residents should be chosen by a table of random numbers. And, of course, the residents will be assured of confidentiality. Interviewing people for the purpose of obtaining personal feelings and experiences requires skill and sensitivity regardless of whether the person is aged or not. A recent experience is relevant here.

A medical student was conducting a study about how and why aged people chose one or another prescription benefit plan from the many contained in the recent Medicare legislation. There was much in the media about the long list of options seniors had to consider. He asked for and received permission to interview residents of the center I reside in. Participation was voluntary. I volunteered even though I did not change my prescription plan which I long had.

It turned out that I would not be available on the 2 days he was here. One of the residents who was interviewed told me that she was given a test of memory: She was told that she would hear a list of words and later in the interview would be asked to recall as many on that list as possible.

I was furious. For one thing, no one had been told they would be given a test of memory. For another thing, it would be obvious to residents of this center that if they had difficulty choosing a plan and then got a low score on the memory test, it would be interpreted as a sign of mental decline. Even before Alzheimer's disease became household words, aged people panicked when they had difficulty remembering this or that. And why not? On TV or in movies or in novels, or the jokes told by comedians—how can one avoid the stereotype of the forgetful old person? But when we are young the stereotype of the aged person has no unusual relevance to us. But once we go from 60 to 70 to 80 and thoughts of bodily decline occupy our thoughts, errors of memory take on an ominous meaning. If there is anything a resident fears, it is far less that their memory is not what it was but that it is getting worse.

Anxiety and stress can have adverse consequences on memory, especially when we are brought up to accept the stereotype of the forgetful, aged person. I am in no way denying that memory loss in aged people may be and is frequently a manifestation of what some regard as "normal senility." There is a world of difference between blithely using a low test score on a memory test as proof positive of a brain change and overlooking alternative explanations. That is why I referred the reader to Koch's work with nursing home patients, and Schaefer-Simmern's work with long institutionalized, mentally regarded individuals.

Why is it that some permanent residents of the center where I reside have problems recalling when or if a family member visited them in the previous week, or whether today is Tuesday or Sunday, or if yesterday was the occasion of a storm, or a clear bright day? These tend to be people for whom each passing day is in no way different in routinized experience. They spend their days inside the facility and there are no markers of time that have personal meaning for them. Why should one expect them to look forward to tomorrow? Their world is a shrinking one.

Many of the aged have concluded that they cannot do what they would want to do or learn. They get through their days in various ways, resigned to the conclusion that their best days are far behind them and fear to put into words to others what they would want to learn and do. But there is a group, smaller to be sure, of those who at a very advanced age continue to do what they have long done. Some engage in an activity they long wanted to engage in but for one or another reason did not or could not do so. Let me illustrate this by discussing two people who are in their nineties, one a resident of the center in which I reside, and the other lives in the community.

Maria Levinson has been in the Center for 5 years. Before she came here she was very active in a variety of organizations having to do with women's issues, civil rights, the UN. Since coming to the Center she has initiated or participated in book clubs, discussion groups, and more. But there is one activity program she initiated which best suits my purpose in this chapter: to remind the reader that what people are and can become depends on opportunities to engage in something new and challenging to them and gives them the feeling they are worthy.

Yale has approximately 3000 foreign students who will be there for 1–4 years. They are mostly postdoctoral students in specialized fields of medicine, computers, etc. Yale has long had an office specifically for these foreign scholars to help them in whatever problems of adjustment they may have. One of their problems is the realization that their clarity of spoken English is not as good as they expected. An effort is then made to find someone who will try to be of help to them. Well, one day Maria called the Yale office and said that there were residents of her center whom she was sure could be able to be of help. Why not use them? Maria would screen them and match each of them with a suitable resident. The Yale office was more than willing to explore a new source of helpers they very much needed. The result has been that at any one time there are about 20 residents who are working with one or more of the foreign scholars. What is truly heartwarming is the number of friendships which have developed between the helpers and the helped. And Maria is the organizing center which keeps the program going to the delight and stimulation of everyone, including the Yale office for foreign students.[1]

The second example is about someone, a friend, who is not a resident of the Center but I include him for reasons which will become clear shortly. His name is Walter Jacobson who is in his mid-nineties. Walter was a Yalie and a highly competent engineer. When he reached retirement age and had to retire, he was not the kind of person who would allow himself intellectually to go to seed, there was too much he wanted to know in fields of study about which he was curious. He lived in New Haven in walking distance of a new campus of a state university. He learned that senior citizens could enroll in a course as long as enrollment for it did not prevent a "regular" student from taking it. There was no fee. For Walter this was like manna from heaven. Over a period of about 25 years he has taken every course he had ever wanted to take. As one might have expected, his presence in the courses was warmly appreciated by the instructor and the other students.

What is common to each of these examples? One is that in both instances an opportunity was presented for aged people to do something they wanted to do. In one instance the opportunity was created by Maria, in the other it was the state college. I am not aware of any resident of the Center who has taken a course at the college. I have no doubt that if the president of the college had come to the Center and "invited" residents to consider taking courses, there would have been more than a handful of takers. That was essentially what Maria did. She made it easy for residents to say yes or no. She assured them of whatever support she could be. When above I used the term opportunity, I was referring to an interpersonal context in which the opportunity is discussed with no concern for differences in age or status. We live from an early age in a society in which you cannot avoid hearing that old people's cognitive capacities deteriorate, particularly memory. (It deserves emphasis that it was Maria who approached the Yale office for foreign students not vice versa. Community agencies do not think of centers for endings as a resource that may be useful to them. No one at Yale-New Haven Hospital asked the members of the "Knit and Stitch Group" to make clothes for newborns of poor dependent mothers. The initiative was that of a member of the group.)

When David Wechsler decades ago standardized his new intelligence test for adults, he presented a graph demonstrating that around 50–60 years of age the average IQ begins to decrease. It never occurred to Wechsler that 60-year-old people may, in contrast to younger people, experience the test situation with more anxiety and self-doubt, if only because it has been decades since they took such a test.

I am in no way suggesting that Wechsler's concept of normal deterioration has no validity. I am suggesting that one should not automatically assume that his or her

---

[1] A month after this chapter was written, there was a brunch at the Center for the scholars and the residents who, like me, served as "conversationalists." Each of the scholars and their wives brought a special native dish. There was a Chinese singing group. Twenty five people attended.

score does not mean that culturally learned attitudes about being old can be ignored or glossed over.

No one had to tell Walter Jacobson that the college presented an opportunity for which he was internally primed. But in the case of Maria's project, the residents were not at all primed immediately to say yes. They needed a Maria to describe the opportunity which had never occurred to any of them. She offered her personal support with the implicit hope they would think it over.

It is unfortunate that in his book about his experience in poetry writing by old people in a nursing home, Kenneth Koch says very little about the way he got them to want to engage in the project. We can assume that he did not walk in, make an introductory speech, and the old people went from apathetic to eagerness. Engendering in others the feeling *to want* to have a new experience is no routine affair.

It is beyond the scope of this book to expand on why we know so little about the factors which play a role in the style of adjustment to institutional living. What I wanted to convey is the caveat that the commission not gloss over the tendency of personnel to ignore what is external to the resident: The degree to which their perception and judgment of an individual is related to the ambiance, programs, or lack of programs, or the willingness and ability of staff to view the individual other than in terms of stereotypes of what aged people are and can become. Stereotypes short-circuit thinking and reinforce routinized, automatic actions.

We are now used to hearing that in the coming decades, medical science will have found means to lengthen the life span. I do not react to that prediction with undiluted enthusiasm. Medical science sees its goals as a boon to human existence, but medical science has little or nothing to say about the quality of that existence. It is considered legitimate to support the efforts of medical scientists. But let us not forget that it is no less legitimate to ask: What do we owe aged people that would cause them to want to see tomorrow? We nod immediate assent when we say we "owe" to infants the attention and support to keep them alive. We owe the same to aged people who are becoming increasingly dependent on others who want them to see tomorrow. Infants have no language to express needs and wants. Aged people have the necessary language but far too many caretakers have ears that do not hear. Our internal voices remain internal rather than expressed when we learn that no one is capable of hearing us. The aged person retreats in silence.

Chapter 9
On the Uses of History

In my initial outline for this book I planned to devote the opening chapter to times in our national history when we failed or were unable to confront a future problem already waiting to enter the stage of history. The failure or inability to confront these problems set the stage for mammoth changes in our society at the cost of human and material sources. The more I pondered that outline the more I became convinced that such a history, however brief, would be more appropriate and more impactful if it came after a description and discussion of a current problem like the care of aged people. After 45 years of teaching university students I knew that history qua history has to have personal relevance. When Henry Ford said that "history is bunk," he was unaware that he was saying that people in the present were different than people in the past, a belief for which there is no evidence whatsoever. In the present as in the past people were thinking, feeling, acting, planning, organisms capable of clear vision and blindness, creativity and stupidity, reason and self-defeating passion, and as the king of Siam was wont to say in a Broadway musical: etcetera, etcetera, etcetera. It was inconceivable to Henry Ford that we are kin to those who lived in the near and long past. History does not repeat itself, people do, and that is the fascination and contribution of history. And that is why Santayana said that those who ignore history are doomed to repeat it. Henry Ford did not live long enough to know that the twentieth century was the most murderous in recorded history.

The first example concerns slavery in America. Before slavery was legitimated in the constitution, there were those who viewed it as abomination which God would punish, and there were those who justified enslavement of blacks as discharging a God given obligation to care and rule over a less than human people. It is not projecting present onto the past to say that the opposing camps believed that their differences were irreconcilable and would lead to catastrophe. It is true that at the 1787 constitutional convention both groups knew that if slavery was not legitimated, there would be no United States. In the decades following the convention the moral issue became more heated and bitter, culminating in the bombing of Fort Sumter by the South. We live with the consequences today and we will live with them for decades to come.

Thousands of books, monographs, and articles have been written about the origins of the Civil War and it is hard to avoid concluding that the war was inevitable. But in the realm of human affairs what we say is inevitable does not rule out the

S.B. Sarason, *Centers for Ending*, Caregiving: Research, Practice, Policy,
DOI 10.1007/978-1-4419-5725-2_9, © Springer Science+Business Media, LLC 2011

role of contingency. So, for example, we—in hindsight of course—imagine scenarios which may have cast the slavery issue in a radically different light: Thomas Jefferson seriously considered sending slaves back to Africa. That and other suggestions fell on deaf ears, which got deafer and deafer with each passing year. Northern leaders offered compromises but the South was not disposed to compromise. And the North would not desist in their efforts to limit and ultimately to eliminate slavery. The stakes in their zero sum game got higher and higher. However I cannot refrain from noting that one of the founding fathers was a slaveholder and opposed to any equality between blacks and whites. Not long after the United States was formed he wrote words expressing his cataclysmic foreboding of what he feared might happen.

There must doubtless be an unhappy influence on the manners of our people produced by the existence of slavery among us. The whole commerce between master and slave is a perpetual exercise of the most boisterous passions, the most unremitting despotism on the one part, and degrading submissions on the other. Our children see this and learn to imitate it. . .. The parent storms, the child looks on, catches the lineaments of wrath, puts on the same airs in the circle of smaller slaves, gives a loose to the worst of passions, and thus nursed, educated, and daily exercised in tyranny, cannot but be stamped by it with odious peculiarities. The man must be a prodigy who can retain his manners and morals undepraved by such circumstances. And with what execration should the statesman be loaded who, permitting one half the citizens to trample on the rights of the other, transforms those into despots, and those into enemies, destroys the morals of the one part and the amor patriae of the other. . .. With the morals of the people, their industry is also destroyed. For in a warm climate no man will labor for himself who can make another labor for him. . .. And can the liberties of a nation be thought secure when we have removed their only firm basis, a conviction in the minds of the people that these liberties are of the gift of God? That they are not to be violated but with His wrath? Indeed I tremble for my country when I reflect that God is just; that his justice cannot sleep forever; that considering numbers, nature and natural means only, a revolution of the wheel of fortune, an exchange of situation, is among possible events; that it may become probable by supernatural interference. The Almighty has no attribute which can take side with us in such a contest [p.56].

Those are the words of Thomas Jefferson. They are taken from Commager's book (1975) *Jefferson, Nationalism, and the Enlightenment*.

The Northern and Southern leaders were highly educated and historically minded men. Let us imagine that we meet with them a few years before the civil war started and we say to them: "You are on the path to war and you are ignoring something you know about war in the near and long past. What you know and are ignoring is that war changes everything and everyone. Both sides will find that what is called victory is a bitter cup of tea, a cause of regret, of unwanted upheavals, of new and unpredictable and insoluble problems. In short, you are ignoring the obvious: As victors or losers you will find yourselves a very pathetic lot. You are ignoring the basics but your passions and arrogance prevent you from recognizing it."

At the 1787 convention the colonists did not raise or even imagine that a future civil war was a possibility. In principle that is what I have argued about America's current failure to confront the societal transformation that will be a consequence of the huge increase of aged people who will require care and facilities for which there is no planning, let alone discussion. *Unlike the colonists, however, the coming societal transformation is a statistical certainty, not a figment of a fevered imagination.* And when that transformation will begin to seep into the consciousness of American society no future historian will be able to say that what happened was unknowable or unimaginable.

But what about the role of contingency? Is it not possible things might happen which will mightily reduce or abort the worst features of the transformation? The answer is yes but I find it hard to imagine that such contingencies will not be instances of cures that are worse than the disease. And nothing is more illustrative of the potentiality of the unpredictable consequences of contingency than what happened once the Civil War ended.

Would it have been a difference that makes a big difference if Lincoln had not been assassinated? Bear in mind that more soldiers were killed in the Civil War than in World War II; the peace terms at Appomattox were generous in the extreme; neither his Gettysburg Address in 1864 nor his second inaugural address in 1865 contained a denunciation, or fulmination against the South. I have never heard or read a historian of Lincoln who disagreed that the post Civil War decades would have been mightily less difficult had Lincoln finished his second term. Planning for the future assumes that leadership will be continuous and able to deal with unpredictable contingencies. When that leadership suddenly changes by an assassin's bullet, all bets are off, the future is too cloudy. That is what happened when President Franklin Roosevelt died in Warm Springs, Georgia. Who was the Harry Truman who was sworn in? He was a relative unknown and untested figure who seemed to be a nonentity compared to the late president. No one, but no one, expected that future historians would come to rate Truman as one of our great presidents. By definition, contingency only tells us that our future has changed.

Unfortunately, in regard to issues of aging I have discussed in this book, there has been no leadership whatsoever. We are drifting to a very problematic future. We are, so to speak, flying blind with no instruments to warn us that we are ignoring data that tell us that there is grave trouble ahead which will transform America as we now know it. So let us turn to another example where national leaders of different countries planned a future that ignored everything that had been learned about human behavior and as a result set the stage for the most destructive century in human history. What made these leaders ignore what had been written large in human history?

I refer to the peace conference that followed the end of World War I. The primary aim of the conference was to insure that never again would Germany be a threat to Europe. To make a long story short, it did so by requiring Germany to pay an amount of reparations that was utterly unrealistic and self-defeating of all of Europe. Revenge was the motivating factor for what went on at the peace conference. Greed was also a factor. Versailles was no Appomattox. It had no intention to bind wounds, it wanted a nation to suffer the indignities of defeat, powerlessness, and poverty. I am sure that every leader at the conference was well acquainted with what Voltaire had said early in the previous century: History is written by the victors. What the leaders ignored was that Voltaire's quip was an indictment of victors unwilling or unable to include their mistakes and frailties as causative factors in the histories they wrote. Versailles set the stage for World War II at the end of which some, and only some, of the worst mistakes of Versailles were not repeated.

The final example is one about which all adult readers will, to some extent, be familiar. It concerns a type of economic occurrence that has marked our national history. I refer to the fact that periodically the country experiences a depression, a recession, a panic which has catastrophic consequences in the lives of many of its citizens. Why did it take a hundred and fifty years to take actions to prevent or dilute the strength of such occurrences? The brief answer is a combination of such factors: hope springs eternal, short memory, and a socioeconomic system that puts a premium on a rugged individualism in the pursuit of money and power. These are words being written by someone well aware of the virtues of the capitalist system. But I am also well aware that it is a system that contains dangerous self-defeating features. That is not a judgment but a fact, just as it is a fact that we are at the beginning of a demographic revolution which will transform our society. I use the word transform in a neutral descriptive sense. I have no doubt, however, that it will have both positive and negative consequences, that there will be divisive controversy about costs, it will require changes in how health personnel will be selected and trained, and it will expose the gulf in services for the haves and have-nots, and it will confirm the law that problem solution begets problem creation. The crucial positive consequences are having done for others what we will want to have others do for us if and when we will need care. Care does not mean being merely kept alive but that one is encouraged and stimulated to know that he/she is not alone or lonely, that he/she is worthy of other people's respect and their desire to help. One looks forward to being alive tomorrow. We come into this world wired, so to speak, to need air, food, and touch. In a matter of months the indissoluble relation between touch and love begins to form, its vicissitudes are many, its quality and frequency will vary, and it is a very complex affair both for child and parent. But one thing we learn: we are needed by others and we need each other. In the final years of our lives both needs take on saliency and strength. That is not easy for younger people to understand. That is why in earlier pages I said that care and understanding of aged people is first a moral obligation that facilitates but does not guarantee that psychological understanding will follow. Humans are not "wired" to make understanding easy.

I make no claim that this book cornered the market on truth in regard to the issues I have posed and the suggestions I put forth. I am certain, however, that no knowledgeable person will deny that we are and will increasingly be confronted with the implications of a very changed demographic composition of our society. What is so worrisome (at least to me) is that the issues have received no serious public attention. Beginning with the presidential campaign of 2000 I have watched more speeches, debates, and symposia than I can count. What I can count is the number of times a political and/or public figure discussed the implications of the demographic projections. By my count, the number was zero. They did discuss it ad nauseam in regard to what it will mean for the funding of social security, that the baby boomers would be shortchanged, that they already are reluctantly and angrily resigning themselves to significantly lower social security payments. I discussed what I was writing with an economist colleague, who said he could not quarrel with the implications I had drawn from the demographic data. He went on to say something like this: "How can you expect political leaders to talk about implications? We have a humongous national debt, we run large annual budget deficits, our international trade deficit is very large and we are in a war in Iraq. Where will the money come from for what you are advocating? Nobody wants to talk about it. It's not a vote getter."

Total silence is no answer, just as it was for years no answer to the *possibility* that global warming may be no transient phenomenon. Nations cannot afford to play Russian Roulette. A nation's leaders have the obligation to confront problems, not to bury them. It should go without saying that developed nations have a surfeit of problems, but not the resources to do justice to all of them. We may be clear about what we believe to be the ideal solution for any of these problems, but we also have to know that we will always fall short of the ideal. And how far we fall short of the mark, the gulf between the ideal and the doable will depend on, among other things, how soon we recognize and confront the problem. When AIDS first appeared on the medical-social radar screen, it was not recognized as a disease that could and would become a world-wide catastrophe. But once it was recognized for what it is, actions (research, education) quickly followed. No one can deny that many lives have been saved, but neither can anyone deny that the problem has not been "solved" to the degree we desire. The major human social problems do not have "solutions" in the sense that four divided by two is two is a solution.

What I have written about the phenomenology of aged people is not understandable to people who do not see themselves as aged. A person of fifty may have aged parents and has gained some understanding of how they see themselves and their future. The degree of understanding will be a function of his or her long relationship with the parents, the frequency with which they have interacted, the quality and closeness of their relationship, and their ability or willingness to reveal themselves to each other. Self-revelation of one's fears, conflicts, and expectations does not come easy to people, as every psychotherapist will attest. Far more often than not, aged parents and their offspring share the belief that there is little to be gained by talking about "sensitive" issues which have no solution. Who wants to talk about how someone is coping with a fore-shortened future? To say that you understand someone is to imply that you now know something you did not know before.

In the 1960s, blacks made it crystal clear that they believed that whites could not understand them. What blacks failed to see is that whites have difficulty understanding whites. Whatever I have said about understanding is a predictable feature of all human relationships. It is especially difficult in the case of aged people who vainly try not to think about what their final years may bring them.

In the beginning of this book I explained why the initial chapters were going to be very personal, not primarily to be self-revealing but to try to convey to others how my suffering went unrecognized or made more painful by the insensitivities of those responsible for my care. It was no source of balm to know that these personnel were doing what they were trained to do, and I refer here to the top administrators, the very part-time physician, the nurses, and the aides.

Dr. Burton Blatt, an educator who could not countenance injustice and insensitivity once said to me, "The inadequacies of our schools of education is in part due to the fact that we say we are training teachers. You train dogs, you educate teachers. More than anyone in this country, Dr. Blatt's writings, especially his book *Christmas in Purgatory* (1970) was largely responsible for the demise of the warehousing of mentally retarded individuals. Unfortunately his premature death did not allow him to know the impact of his writing and actions. Dr. Blatt recognized early on that the education and training of the medical community and its allied workers had all of the inadequacies of the educating and training of teachers.

Epilogue

In February 2008 when I finished writing this book, my attention was caught by an article about a recent study from the Center for Disease Control on suicide rates for different age cohorts in the population. The point of the article was that the rate of suicide for people in midlife has unexpectedly and dramatically increased. That, of course, would be one of the post World War II baby-boom generations. It was after World War II that the phenomena of midlife crises gained currency to a degree it had not had before. The midlife crisis is psychologically a review of one's past and a decision that will lead to a more satisfying future. In the pages of this epilogue, I speculate about the attitudes of baby boomers as they encounter the quality and costs of protective environments many of them will need.

In 1977, I wrote a book *Work, Aging and Social Change*. Its subtitle was *Professionals and the One Life One Career Imperative*. It was an unfortunate title because it conveyed the impression that the book was about older people. In fact, the first two-thirds of the book is based on interviews with college seniors who are making a choice of career; in addition, using 3 years of Who's Who in America, we determined the evidence of career change. Let me explain why we focused on career choices and changes.

Up until World War II, choice of career was generally assumed to be governed by what I called the one life one career imperative. That is to say, you can be A or B in life but you cannot be both, you have to make a choice. It was in all respects similar to the expectation that when you married, you would stay married to the spouse you chose, an expectation based on religious belief and given practical support by state legislation which made obtaining a divorce very difficult, costly, and time-consuming.

The social change did not begin after the war but during the war. There were jobs aplenty for women who had never worked, except of course as housekeeper and the rearing of a family. Men in the Great Depression who had been unemployed or underemployed, or were physically handicapped had no difficulty finding a job. That was even the case for aged people who wanted to contribute to the war effort. It is an incomplete description to say that discrimination in the workplace or on the basis of gender, age, race, and ethnicity was for all practical purposes virtually nil. No less noteworthy was that millions of men and women experienced the sense of self-worth. What about the millions of soldiers in different parts of the world,

S.B. Sarason, *Centers for Ending*, Caregiving: Research, Practice, Policy,
DOI 10.1007/978-1-4419-5725-2, © Springer Science+Business Media, LLC 2011

separated from loved ones, wives and families? The briefest answer is contained in the song sung for soldiers in the musical *South Pacific*: "There is nothing like a dame."

World War II was a long war fought in sites whose attitudes toward sexuality not as Victorian as in America. There is a theme elaborated in my semiautobiographical written after the war, and it is a theme in many of the case histories of servicemen I saw as a consultant in VA clinics and hospitals. The problem was not only the need and pressure for sexual expression but also the guilt and frustration they experienced in adjusting to their wives and friends when they returned home; they were like strangers to each other. But the feeling of estrangement was beyond the sexual to the poignant sense that although the returning GI had changed dramatically, the society had not changed in ways it should have. That theme was well depicted at war's end in the movie *The Best Years of Our Lives*. It is the only movie I have ever seen that ends with an epilogue in which the central male character speaks directly to the audience about why and how the country must change its traditional values about what it owes its less fortunate citizens who for reasons not of their making are denied opportunities to overcome unwelcome disruptions in their lives. It was the theme of social-economic justice. That was a theme during the war publicly and militantly proclaimed by blacks in and out of the armed services who saw the contradiction between fighting in a war against the fascist despots while blacks were segregated in the armed services where they were assigned the most menial duties. It is a little known fact that during the war women who were licensed civilian pilots were not eligible to be part of the Air Force but unofficially were hired to perform mission within the Continental United States. On May 4, 1992, in its series *The American Experience* a documentary *Flying Girls* tells the story of these female pilots.

The reader may be asking: What do these events I have related have to do with nursing homes and aged people in 21st century? My answer is in three parts:

1. If you take seriously that war changes everything and everyone, it is expected that soon after the war ends signs of a social change can be sensed and those initially vague impressions are reflections of what people were thinking, doing, anticipating, and wishing during the war years. How those themes gain shape and force is unpredictable except in board outline. In the case of World War II, every theme I have discussed above came to the front and center in the post war social change. By the 1960s, no one was in doubt that an encompassing social change had taken place. It would take a book of immodest size to describe what had happened. Most relevant for my purposes are the changes which occurred between parents and their baby-boom children.

    The reader should keep in mind that the parents of baby boomers had lived through The Great Depression and World War II. It should occasion no surprise that they greeted the war's end with the attitudes many of them proclaimed. They publicly gave voice to these attitudes in this way: "The world in which we were reared was all screwed up in all kinds of ways. It is not a world we want our children to experience. Our responsibility is to ensure that they will build a better world for themselves. They should have opportunities we never had to become what we wanted to become. They will have more education than we had. They will be smarter and wiser than we were. We owe our children a great deal. If we screwed up, the best thing we can do for them is to help them build a better world than was given us."

The above expression of parental attitudes and goals begs this question: How could or should these goals inform child-rearing practices? These goals were certainly well intentioned, but they required more than inspiring words; they required a practical, concrete set of principles consistent to those goals. It is the old means–ends problem. Different goals, different means.

We have no hard data to answer the question. I know of only one fact which indirectly bears on the issues. The war's end coincided with the publication of Spock's book on infant and child care. It was the first book ever by a pediatrician that told parents they knew more about their child than they thought and they should refrain from slavishly following experts who have rigid rules for rearing children. His book was not technique oriented, but a child-centered one, and that meant that in their interaction with their infant or young child they should be guided by what the parent had experienced and learned about their child.

In the 1960s, when the post World War II social change was clearly per-ceived, Spock was criticized for having advocated "permissiveness" in child rearing which, they said, explained why young people of that era were rebel-lious, destructive of tradition, and self-indulging. That is a confirming instance of Mencken's caveat that for every major problem there is a simple answer that is wrong. Then, as now, the social change remains a mind boggling mystery, especially for people who did not witness or experience those post-war decades. And one of its mysteries is how to explain why for decades the world made the sales of his book exceeded only by the Bible. That brings to mind another of my favorite caveats. It is hard to be completely wrong.

2. Let us turn now to what baby boomers were expressing in the two decades after the war. The first in point of time was the steady increase in the rate of divorces. That suggests there were more than a few fragile marriages which did not end in divorce "for the sake of the children," a justification psychotherapists frequently heard from their patients. That divorce is upsetting to children goes without saying, and the most frequent reaction is a combination of anger, loss, and disillusionment. I do not consider it happanstance that as in the case of the divorce rate the number of children referred to child guidance clinics zoomed upward at a faster rate. It was a harbinger of what was labeled as "the genera-tional gulf." It was in the mid-1950s that the gulf was plain to see. I refer to how in less than a year an unknown Elvis Presley and his rock and roll music became magnets to scores of millions of young people. It was much more than a change in taste, but also an articulate rejection of and a rebellion against the music their parents more than liked sung by Frank Sinatra, Bing Crosby, and Perry Como. What parents judged to be close to barbarian, their children considered to be the dawning of a new age. In less than a few years, the romantic Hollywood musical became a thing of the past. Then came, among other things, the explosion of the

hydrogen bomb, to parents to make their cellars livable in case a bomb exploded; teachers told their students what to do in case of an atomic war; the Soviet Union blockade of Berlin and the fear that World War II would erupt; the Korean War; with the possibility that a war with China would break out; the Supreme Court's desegregation decision and the display on TV making it quite clear, especially in the South, that the virus of racial prejudice could or would become more potent. The 1950s were later called the "silent fifties." No label could be more inappropriate and untrue.

The gulf between baby boomers and their parents was basically cultural in content in that baby boomers were rebelling against a style of living undergirded by standards and morals, the baby boomers considered barriers to personal expression and freedom to decide what you will wear, who your friends will be, what you do with your friends, and not be hemmed in by a boring family life that was dispiriting and the opposite of happy; if they knew what the word dysfunctional meant, they would have used it as a label for how they judged their parents and family life.

Why did Salinger's novel *The Catcher in the Rye* come to be regarded by baby boomers as a realistic and sensitive depiction of their own lives? The central character was Holden Caulfield who felt alone, confused, lost, bored in school, emotionally unconnected with his parents and possesses a strong and ever enlarging fantasy life about the pleasures and freedom he had in Manhattan. His adventures are both poignant and comedic and he ends up more psychologically disturbed than ever. If the book was long a bestseller and its sales still are large, it is despite the fact that since its publication, there have been many school libraries and boards of education who prohibit making the book available to children.

I said that the post World War II social change was a countercultural one. In the decade before the war, there was also a generational gulf, but it was about the political-economic features of a capitalist system responsible for the disaster of the Great Depression. The World War II baby boomers were rebelling against what they regarded as a Puritan-Victorian ethos that was restrictive, repressive, conformist, and infantilizing of its youth at the expense of a personal freedom to decide how one will live and with whom. The baby boomers have been described as the "me generation," a label meant to imply that they were interested in their individual selves and goals. Ironically, the label describes a rugged individualism markedly different than what had been the traditional conception of American rugged individualism.

Finally, I come to the third part of my answer to the question: What is the relevance (if any) of those born early in the baby boom generation – those who grew up in an era of significant social change – experiencing the many inadequacies of nursing homes more than a half-century later? Obviously, the question in no way suggests that the baby boomers caused these inadequacies. What the question has suggested is that we examine how what we know about nursing homes will be experienced and reacted to by what we think we know about baby boomers.

There have always been generational gulfs be they slow, small, and hardly cause for comment. There are gulfs which are divisive, mostly rancorous, and characterize an era such as the consequences of the Civil War, World War I and II. The baby-boom generation after World War II was embedded in, affected by, and ultimately participants in a truly pervasive social change.

In 1987, Alan Bloom's book *The Closing of the American Mind* was published. It received excoriating reviews by proponents of the post World War II change. Despite those reviews, the book was on best seller lists for weeks on end. Bloom judged the social change as a near total cultural disaster in which mindless, rebellious youths and many equally mindless adult supporters became the avant garde in storming the barricades, destroying them, and transforming what was best in American culture. Twenty years after the book was published, the conservative journal the New Criterion devoted one of its issues to a defense of the accuracy of Bloom's analysis and meditations (Kimball, 2007; Pierson, 2007; Steyn, 2007).

There is only one thing about which proponents and opponents were in total agreement: the social change had been a rebellion against all forms of received knowledge, wisdom, and culture.

Question: Is it not reasonable to assume that the millions of baby boomers already knocking on the doors of old age will by no means be clones of those who are aged today? And is it not reasonable to expect that an undetermined but large number of baby boomers will want or need a protective environment? How will they react to a situation in which there is a shortage of nursing homes, assisted living, and total care facilities; and in the case of assisted living and total care facilities, the cost of entering and residing in them are well beyond their financial resources? Will they grin and bear it, take it lying down, keep their disappointment, sadness, and anger bottled up at the awareness that they are not masters of their fate or captains of their souls, that the society they fought to transform, like them as individuals, had not taken the demographics seriously.

Can one dismiss out of hand the possibility that among those baby boomers who rebelled against tradition, customs, and authority, there will be some who will come to regret their actions and goals despite manifestations of success? The reasons may differ from one individual to another. It has almost always been the case, in encompassing social change that some of its proponents become disillusioned, regretful, that their rebellion has not been followed by feelings of satisfaction, and rewarding personal relationships.

Let me illustrate my point by my experience at Yale. Teaching and supervising responsibilities were primarily with graduate students seeking their doctorates. Those responsibilities began in 1945 and ended with my retirement in 1989. I got to know well hundreds of those students from all of whom I learned how and why they embraced some or all of the different aspects of the social change. These students had been admitted on the basis of a clear agreement with the goals of the Yale department: to learn and do research, obtain a university position, and to be productive researchers. My guess is that 15 percent of the students dropped out of the program because it did not have for them relevance in the "real world." Thirty percent met the first two goals, but when they obtained a position in a college or university they published little or nothing but devoted themselves in one or another form of social action; some of the men and women in this group were angry that efforts to better the society were less valued in the university than doing research and some of them felt guilt about having chosen psychology as a career. Approximately half of the students met all the goals of the department and some of them devoted their research career to studying problems directly devoted to issues of social change.

All of the above are surface phenomena. Faculty–students interactions are between two people having very unequal power and status. Matters were not helped any by the fact that many of the students were veterans who were older than I. Every organization, small or large, has a rumor mill and that is how I learned the obvious: students like everyone else are very complex people who will inhibit revealing anything which could be interpreted as critical and rebellious.

Women students were far more eager than the men to volunteer to me the fact that they would not get married and have children until they had a university position, something I usually knew via the rumor mill.

I have often wondered what problems and attitudes change the students (all of them I experienced after they left Yale). I can think of no explanation which predicts that they will experience old age any differently than the generations before them. That is why my attention was directed to a lead article in the New York Times of February 19, 2008 (Cohen, 2008).

A new 5-year analysis of the nation's death rates recently released by the federal Centers for Disease Control and Prevention found that the suicide rate among 45–54-year-olds increased nearly 20 percent from 1999 to 2004, the latest year studied, far outpacing changes in nearly every other age group. (All figures are adjusted for population.)

For women 45–54, the rate leapt 31 percent. "That is certainly a break from trends of the past," said Ann Haas, the research director of the American Foundation for Suicide Prevention.

By contrast, the suicide rate for 15–19-year-olds increased less than 2 percent during that 5-year period—and decreased among people 65 and older.

The question is why. What happened in 1999 that caused the suicide rate to suddenly rise primarily for those in midlife? For health experts, it is like discovering the wreckage of a plan crash without finding the black box that recorded flight data just before the aircraft went down. . ...

Without a "psychological autopsy" into someone's mental health, Dr. Caine said, "We're kind of in the dark."

The lack of concrete research has given rise to all kinds of theories, including a sudden drop in the use of hormone-replacement therapy by menopausal women after health warnings in 2002, higher rates of depression among baby boomers or a simple statistical fluke.

At the moment, the prime suspect is the skyrocketing use—and abuse—of prescription drugs. During the same 5-year period included in the study, there was a staggering increase in the total number of drug overdoses, both intentional and accidental, like the one that killed the 28-year-old actor Heath Ledger. Illicit drugs also increase risky behaviors, C.D.C. officials point out, nothing that users' rates of suicide can be 15–25 times as great as the general population.

The rate of increase in the two cohorts is indeed puzzling. In fact, the rate of increase may be higher than the Center for Disease Control reports because families are reluctant or ashamed to have it known that a family member committed suicide. Neither the CPC nor I know the complicated story of the suicider's life or the circumstance surrounding the suicide. But to someone like me who has a special interest in the baby-boomer generations, I feel justified in speculating—and it is no more than speculation—about the fact that the age cohorts with the highest rate of increase were largely baby boomers born in this country. What we know (or think we know) about suicides is not only that the person feels the future is hopeless, but that it also is a display of anger and rebellion against events in their past and people in their present. The person sees no way out, their misery and despondency have to end, life is not worth living, they cannot forget the past, and they cannot contemplate the future. But those two cohorts are a subgroup in a much larger group of baby boomers in these age cohorts. Does it not suggest the possibility that their judgments about their past and future contain themes of anger and disappointment?

I cannot answer the question and neither can anyone else. It is a question that can only be partly answered by imaginative researchers who can study subgroups in these age cohorts who have not committed suicide. That research is not likely to be carried out. In any event, it is my prediction that the baby boomers who in the next two decades will be part of the group we label aged and will need or want a protective environment which will be in short supply and whose low quality and inadequacies an open secret, will transform a present crisis into a natural, divisive, and moral disaster.

I hope I am wrong. But I have lived through the rise of Hitler's Germany, World War II, Vietnam, the Arab Oil embargo, three Wall Street collapses, the belated recognition of global warming, and the current ever percolating consequences of a collapse of the housing market and a record breaking increase in housing foreclosures. For each of these very complicated occurrences, there were a small number of people who warned against complacency and inaction and denial displayed by political leaders and people generally. The warnings went unheeded. In the case of the care of aged people who will need or want a protective environment, the situation is worse because there is total silence about the "issues" I have raised in this book. If I know that I have not cornered the market on truth, I am convinced by personal experiences and that of other aged people that I am very far from wrong about the major problems I have identified.

About the Author

Seymour B. Sarason was Professor of Psychology Emeritus at Yale University, where he taught from 1945 to 1989. He is the author of more than 40 books and is considered to be one of the most significant researchers in education and educational psychology in the United States. The primary focus of his work was on education reform in the United States. During the 1950s, he and George Mandler initiated the research on test anxiety. He founded the Yale Psycho-Educational Clinic in 1961 and was one of the principal leaders in the community psychology movement.

References

Andres-Hyman, R. C., Strauss, J. S., & Davidson, L. (2007, March). Beyond parallel play: Science befriending the art of method acting to advance healing relationships. *Psychotherapy: Theory, Research, Practice, Training, 44*(1), 78–89.

Bensman, D. (2000). *Central Park East and its graduates*. New York: Teachers College Press.

Biro, D. (2000). *One hundred days: My unexpected journey from doctor to patient*. New York: Pantheon Books.

Blatt, B. L. (1970). *Christmas in purgatory*. Boston: Allyn & Bacon.

Cohen, P. (2008, February 19). Midlife suicide rises, puzzling researchers. *The New York Times*.

Commager, H. S. (1975). *Jefferson, nationalism and the enlightenment*. New York: Braziller.

Consumer Reports. (2006, September). Nursing homes: Business as usual. *Consumer Reports, 71*(9), 38–41.

Gass, T. (2004). *Nobody's home. Candid reflections of a nursing home aide*. Ithaca, NY: Cornell University Press.

Gawande, A. (2007). The way we age now. *The New Yorker*, April 30, 2007.

Ginzberg, E., & Bray, D. (1953). *The uneducated*. New York: Columbia University Press.

Kimball, R. (2007). "Openness" & "The Closing of the American Mind": On the role of ideas of "tolerance" in the intellectual decline. *The New Criterion, 26*, 11–17.

Koch, K. (1970). *Wishes, lies, and dreams: Teaching children to write poetry*. New York: Random House.

Koch, K. (1997, September–October). *I never told anybody: Teaching poetry writing in a nursing home*. New York: Random House.

Kurland, G. (2002). *My own medicine: A doctor's life as a patient*. New York: Times Books.

Levine, E. (2001). *One kid at a time. Big issues from a small school*. New York: Teachers College Press.

Pensack, R. J. (1994). *Raising Lazarus: A Memoir*. New York: GP Putnam's Sons.

Pierson, J. (2007). The closing of the American mind at 20: On Alan Bloom the teacher taking on the listless realms of academe. *The New Criterion, 26*, 4–10.

Rosenbaum, E. E. (1988). *A taste of my own medicine: When the doctor is the patient*. New York: Random House. (Republished in 1991 by Ivy Books as "Doctor" and made into the movie "The Doctor" that year)

Sarason, S. B. (1949). *Psychological problems in mental deficiency*. New York: Harper and Brothers.

Sarason, S. B. (1969). The school culture and processes of change. In Francis Kaplan & Seymour B. Sarason (Eds.), *The psycho-educational clinic: Papers and research studies*.

Sarason, S. B. (1971). *The culture of the school and the problem of change*. Boston: Allyn & Bacon.

Sarason, S. B. (1974). *The psychological sense of community. Prospects for a community psychology*. San Francisco: Jossey-Bass.

Sarason, S. B. (1986). *Caring and compassion in clinical practice*. San Francisco: Jossey-Bass.

Sarason, S. B. (1988). *The making of an American psychologist. An Autobiography*. San Francisco: Jossey-Bass.

Sarason, S. B. (1990). *The predictable failure of educational reform*. San Francisco: Jossey-Bass.

Sarason, S. B. (1992). *The case for change. Rethinking the preparation of educators*. San Francisco: Jossey-Bass.

Sarason, S. B. (1995). *Parental involvement and the political principle: Why the existing governing structure of schools should be abolished*. San Francisco: Jossey-Bass.

Sarason, S. B. (1996). *Revisiting the culture of the school and the problem of change*. New York: Teachers College Press. (Originally published in 1971, 2nd ed., 1982)

Sarason, S. B. (2004). *And what do you mean by learning?* Portsmouth, NH: Heinemann.

Schaefer-Simmern, H. (1948). *The unfolding of artistic activity*. Berkeley, CA: University of California.

Smith, H. (2006, 10 November). *Retirement and the 401 k benefits*. Public Broadcasting System.

Spitz, R. A. (1945). Hospitalism: An inquiry into the genesis of psychiatric conditions in early childhood. *Psychoanalytic Study of the Child, 1*, 53–74.

Spitz, R. A., & Wolff, K. M. (1946). Anaclitic depression. *Psychoanalytic Study of the Child, 2*, 313–342.

Steyn, M. (2007). Twenty years ago today: On rock music's oppressive rule over society. *The New Criterion, 26*, 18–23.

Towbin, A. (1978, Winter). The confiding relationship: A new paradigm. *Psychotherapy: Theory, Research and Practice, 45*(4), 333–343.

Weisman, J. (2002). *As I live and breathe: Notes of a patient-doctor*. New York: North Point Press.

Index